NITERNATIONAL SERIES OF MONOGRAPHS IN

NATURAL PHILOSOPHY

GENERAL EDITOR: D. TER HAAR

VOLUME 16

SPECTRAL PROPERTIES OF DISORDERED CHAINS AND LATTICES

OTHER TITLES IN THE SERIES IN NATURAL PHILOSOPHY

Vol. 1. DAVYDOV—Quantum Mechanics

Vol. 2. FOKKER—Time and Space, Weight and Inertia

Vol. 3. KAPLAN—Interstellar Gas Dynamics

Vol. 4. ABRIKOSOV, GOR'KOV and DZYALOSHINSKII—Quantum Field Theoretical Methods in Statistical Physics

Vol. 5. OKUN'—Weak Interaction of Elementary Particles

Vol. 6. SHKLOVSKII—Physics of the Solar Corona

Vol. 7. AKHIEZER *et al.*—Collective Oscillations in a Plasma

Vol. 8. KIRZHNITS—Field Theoretical Methods in Many-body Systems

Vol. 9. KLIMONTOVICH—The Statistical Theory of Non-equilibrium Processes in a Plasma

Vol. 10. KURTH—Introduction to Stellar Statistics

Vol. 11. CHALMERS—Atmospheric Electricity (2nd edition)

Vol. 12. RENNER—Current Algebras and their Applications

Vol. 13. FAIN and KHANIN—Quantum Electronics, Vol. 1. Basic Theory

Vol. 14. FAIN and KHANIN—Quantum Electronics, Vol. 2. Maser Amplifiers and Oscillators

Vol. 15. MARCH—Liquid Metals

SPECTRAL PROPERTIES OF DISORDERED CHAINS AND LATTICES

J. HORI

Professor in Physics,
Department of Physics,
Hokkaido University,
Sapporo, Japan

THE QUEEN'S AWARD TO INDUSTRY 1966

PERGAMON PRESS

OXFORD · LONDON · EDINBURGH · NEW YORK

TORONTO · SYDNEY · PARIS · BRAUNSCHWEIG

Pergamon Press Ltd., Headington Hill Hall, Oxford
4 & 5 Fitzroy Square, London W.1

Pergamon Press (Scotland) Ltd., 2 & 3 Teviot Place, Edinburgh 1

Pergamon Press Inc., 44–01 21st Street, Long Island City, New York 11101

Pergamon of Canada Ltd., 207 Queen's Quay West, Toronto 1
Pergamon Press (Aust.) Pty. Ltd., 19a Boundary Street, Rushcutters Bay, N.S.W. 2011. Australia

Pergamon Press S.A.R.L., 24 rue des Écoles, Paris 5e

Vieweg & Sohn GmbH, Burgplatz 1, Braunschweig

First edition 1968

Library of Congress Catalog Card No. 66–30636

08 012359 7

Contents

Contents

Preface

As in other fields of theoretical physics, there are two main types of work in the research of disordered lattices. One tries to treat the simplified model as exactly as possible, and the other attempts to construct an approximate theory for the more realistic model by largely relying upon physical intuition. Approximate methods such as perturbation theory and perturbation expansion of the Green function have been successful in many branches of physics, so that it is natural that many investigators have attempted to attack the problem of disordered systems by these methods. It has been proved, however, that approximate methods cannot in general give useful informations about the detailed structure of the spectrum of such a system, being sometimes even misleading, although they may be effective in calculating macroscopic physical quantities. This recognition has been brought about by the recent development of exact theories and the numerical calculations based upon them, which disclosed the fact that the spectrum of a disordered system may have several remarkable properties which are wholly unexpected from the approximate theories.

Apart from such a situation, investigations of the first type are indispensable for clarifying the domain of applicability of approximate theories and elucidating the nature of approximation involved in them, thus giving these theories a firmer basis.

As for the spectral properties of the simplest system such as the lattice with nearest-neighbour harmonic interactions and the corresponding electronic system, the exact theories have now been developed to such an extent that a systematic formulation can be given, which is called *phase theory*. It explains or predicts several characteristic spectral properties such as peak structures of the spectrum, special frequencies or energies, and the localization of eigenfunctions, from one and the same principle. Moreover, it looks out over various systems, which are apparently very different from one another, from a unified point of view without discriminating either vibrational systems from electronic ones, or microscopic systems from macroscopic ones. The present monograph is intended to give a full account of the phase theory.

I wish to thank the members of two series of symposia held with the aid of the Research Institute for Fundamental Physics, Kyoto University, the Symposia on the Theory of Lattice Vibrations of Imperfect Crystals during 1960–2, and the Symposia on the Dynamics and Statistics of Coupled Oscillators during 1963–5, for stimulating the development of the phase theory. I am particularly indebted to Professor H. Matsuda for reading a part of the manuscript and giving valuable advice, to Dr. T. Asahi for keen discussions concerning the fundamentals of the theory, and to Professors M. Toda, T. Toya, and E. Teramoto for continual interest and encouragement. I also wish to express my appreciation of the consideration given to this work by Dr. D. ter Haar, Editor of this series of monographs.

Sapporo,
August 1966 J. Hori

List of Symbols

$\mathbf{a}_k$	An element of a continued fraction
$\mathbf{A}_n$	An element of a dynamical matrix
$\mathbf{b}_k$	An element of a continued fraction
$\mathbf{B}_n$	An element of a dynamical matrix
$c_n^{(\mu)}$	A coefficient of eigenfunction expansion
$\mathbf{C}_n$	An element of a secular matrix
$D(\mu)$	Spectral density
$D(\omega^2)$	Density of the eigenfrequency spectrum
$D_\mu(\omega^2)$	The proper self-energy part
$\mathbf{D}_n$	An element of a secular matrix
$\boldsymbol{f}$	Momentum of an electron
$G(\boldsymbol{r}, \boldsymbol{r}')$	Green function
$G(\boldsymbol{f})$	Fourier transform of the Green function
$\mathbf{G}(\lambda)$	Green operator
$G_{nl}(\lambda)$	An element of a Green operator
$\mathbf{I}$	Unit matrix
$K_{nm;n'm'}$	Central force constant between the atoms at (n, m) and (n', m')
$K'_{nm;n'm'}$	Non-central force constant between the atoms at (n, m) and (n', m')
$\mathbf{K}_n$	A diagonal matrix appearing in a secular matrix
$\mathbf{K}'_n$	A matrix appearing in a secular matrix
$\mathbf{L}_n$ or $\mathbf{L}$	A regular submatrix
$\mathbf{L}^{(j)}$	Impurity submatrix
$\mathbf{L}'$	Impurity submatrix associated with a line impurity
$\mathbf{L}(\lambda)$	A triangular matrix
$\mathbf{L}_n$	An element of a triangular matrix
$M(\lambda)$	Integrated density
$\mathbf{M}_n$	A diagonal matrix appearing in a secular matrix
$\Delta \boldsymbol{M}^{(j)}$	The difference between impurity and regular submatrices
$N(\mathbf{x})$	The norm of a vector $\mathbf{x}$

List of Symbols

$\mathbf{N}$	The total number of eigenvalues
$\mathbf{P}_n$	The dipole operator of the nth monomer
$\mathbf{P}_{n\alpha 0}$	A matrix element of $\mathbf{P}_n$
$\mathbf{P}$	A dynamical matrix
$\mathbf{Q}_n$ or $\mathbf{Q}$	A general or complex transfer matrix
$\mathbf{Q}^{(b)}$	The transfer matrix at the basic atom
$\mathbf{Q}_{n;l}$	A product of transfer matrices
$\boldsymbol{R}_{nn'}$	The vector from monomer n to monomer n'
$\mathbf{S}(\lambda)$	A secular matrix
$t_n(\boldsymbol{r}, \boldsymbol{r}')$	The T-function for a single potential
$T(\boldsymbol{r}, \boldsymbol{r}')$	The T-function
$T(\boldsymbol{f})$	The Fourier transform of $T(\boldsymbol{r}, \boldsymbol{r}')$
$\mathbf{T}_n$	A real transfer matrix
$\mathbf{T}^{(b)}$	The transfer matrix at the basic atom
$\mathbf{T}_{n;l}$	A product of transfer matrices
$\mathbf{U}(\lambda)$	A triangular matrix
$\mathbf{U}_n$	An element of a triangular matrix
$\mathbf{U}_n$	A state-ratio matrix
$\mathbf{v}_n$	The variable of the general vector-difference equation
$\mathbf{V}_n$	The matrix variable of the matrix-difference equation
$\mathbf{w}^{(\mu)}$	A normalized eigenvector of the dynamical matrix
$\mathbf{w}_n$	The variable of a vector-difference equation
$\mathbf{X}_n$	The state vector
$\mathbf{X}_n^{(i)}$	A component of state-vector $\mathbf{X}$
$\mathbf{X}_{\pm}$	The eigenvectors of a transfer matrix
$\mathbf{Y}$	The matrix of transformation of the representation
$\mathbf{Z}_n$	A complex state ratio
$Z_m^{(\mu)}$	An eigenfunction of a regular submatrix
$\boldsymbol{\alpha}_k$	An element of a product of transfer matrices
$\beta_j^{(i)}$	The value of wave-number parameter corresponding to the impurity frequency of the island of the ith kind of the jth impurity
β_μ	A wave-number parameter for a two-dimensional system
$\beta_{\mu\nu}$	A wave-number parameter for a three-dimensional system
$\boldsymbol{\beta}_k$	An element of a product of transfer matrices
$\boldsymbol{\gamma}_k$	An element of a product of transfer matrices
Γ	A finite number appearing in the Green function
Γ_μ	The imaginary part of the pole of $G_\mu(\omega^2)$
$\Gamma(t)$	The characteristic function
$\boldsymbol{\delta}_k$	An element of a product of transfer matrices

$\delta_{\pm}^{(i)}$	Limit phases of the *i*th transfer matrix
$\Delta_{\pm}$	The width of a narrow interval at $\delta_{\pm}$
$\boldsymbol{\Delta}_n$	The phase matrix
Δ_{μ}	Shift of the real part of the pole of $G_{\mu}(\omega^2)$
ε_{μ}	The imaginary part of a complex wave-number parameter
$\varepsilon_{\mu\nu}$	The imaginary part of a complex wave-number parameter
ζ	A critical value for $l_{\min}$
$\varkappa$	The imaginary value of $k = E^{1/2}$ divided by i
λ	The energy parameter
λ_{μ}	An eigenvalue of the energy parameter
λ_{imp}	The eigenvalue of the energy parameter corresponding to an impurity mode
μ	Phonon momentum
μ_k	The *k*th moment of the spectrum
ν	The energy parameter defined by $= 2k/V^{(A)}$; the parameter specifying the boundary condition
ϱ_k	An element of a continued fraction
$\varrho(\boldsymbol{f}, E)$	The probability density in the space of energy and momentum
$\sigma_c^{\pm}(s, \beta)$	Functions giving the critical mass parameter for special frequencies
$\tau_{\pm}(s, \beta)$	Functions giving the critical mass parameter for special frequencies
$\boldsymbol{\theta}_{\pm}$ or $\theta_{\pm}$	Eigenvalues of a transfer matrix
Φ_{α}	An excited state of the total system
χ_{μ}	An eigenvalue of the regular submatrix
$\psi_{n\alpha}$	The state α of the *n*th molecule
$\Psi_{n\alpha}$	A state in which the *n*th monomer is excited
ω_{μ}	An eigenfrequency of a one-dimensional chain
$\omega_{\max}$	The maximum eigenfrequency
$\omega_{\mu\nu}$	An eigenfrequency of a two-dimensional system
Ω	The parameter defined by $2(1 + \gamma) - m\omega^2/K$
$\langle\rangle$	The average; the bound of a matrix
$\{\}$	An interval spanned by limit phases or limit points
$(\mid\mid)$	A matrix element

Introduction

INASMUCH as almost all the solid materials which we encounter in our everyday life are imperfect crystals containing defects and impurities, it is beyond question that the study of such systems is an important subject of theoretical physics. It is of practical interest to investigate how the physical quantities such as electrical conductivity, heat capacity, thermal conductivity, and so on, change when imperfection is introduced. Liquids may be considered as imperfect (molten) solids, so it is also of much interest to discuss their properties from the same viewpoint.

These physical properties depend upon the electronic energy, vibrational frequency, and other excitation spectra of the system under consideration, so that it is important before everything to explore how these spectra are affected by the presence of imperfection. For example, the electrical conductivity of a liquid is determined essentially by whether a spectral gap persists or not when the crystal is molten into liquid. The low-temperature specific heat of an impure solid strongly depends on how its vibrational spectrum is affected by the impurity in the low-frequency region. There is another group of physical phenomena that newly appear when defects are introduced and depend very sensitively upon the change in the spectrum; namely, the extrinsic semiconductivity, impurity-induced optical absorption, resonance scattering by the impurity, and so on.

Theoretically, it is convenient to distinguish two types of imperfect lattices. One is the lattice which contains the defects in a very small concentration, and the other is the disordered lattice which contains a large number of defects or has a liquid-type disorder. The effect of relatively few defects on the spectrum can be investigated very effectively by the Green-function method which was first extensively used by Lifshitz (1942a, b, c) in his pioneering work concerning the vibrational properties of non-ideal crystal.

The existence of localized impurity modes with isolated impurity frequencies was predicted by him for the first time. In 1955 Montroll and Potts independently formulated the method and discussed the properties of impurity modes in detail. Since then, numerous investigations have been made on this subject upon the basis of the Green function.

In recent years, the development of experimental techniques in infrared, γ-ray, and neutron spectroscopy made it possible to observe very clearly the local structure in the spectrum brought about by a small number of impurities, and vigorously stimulated the theoretical study of various impurity-induced dynamic properties of crystals. This in turn is stimulating the investigation of the effect of defects on the electronic, spin wave, and other excitation spectra, which may be treated entirely by the same theoretical method.

Study of the spectra of disordered lattices is, on the contrary, very difficult both theoretically and experimentally. In general, the disorder brings about in the spectrum the features which can neither be obtained by any simple extrapolation from the results obtained for the case of few impurities, nor by highly elaborated versions of the method of the Green function. It is this very difficulty, however, that made the problem an important as well as fascinating one which attracted many mathematically minded physicists, notwithstanding the fact that no experimental data were available which are directly connected with such features.

The first frontal attack on this problem was made by Dyson in 1953, who tried to calculate theoretically the frequency spectrum of one-dimensional glass-like chain. Although he himself could obtain the spectrum only for a very unrealistic model, his method proved to have a profound significance. It implicitly contained a full variety of methods which were later proved to be extremely effective for investigating the disordered lattices (see below). After Dyson's paper appeared, several authors tried to solve the disordered-lattice problem by different methods: variants of Dyson's theory, perturbation expansions based on the Green function, methods of moments and averaged eigenvalue equation, and so on. All of these methods led to the conclusion that the singularities in the spectrum are completely destroyed by disorder, and that it can have only a fairly smooth structure.

This seemingly very natural conclusion was, however, upset in 1960 by Dean. In his epoch-making work, Dean calculated numerically the spectrum of an isotopically disordered linear chain by the method of negative factor counting which is intimately related to Dyson's theory, and showed that it has a very fine structure with many distinct peaks and valleys at the bases of which the spectral density vanishes. Since the publication of this work, a number of computations of the same kind have been carried out. Recently the frequency spectra of disordered two- and three-dimensional lattices have also been calculated and proved to have similar fine structures. Another important fact which was revealed is that the normal modes of disordered lattices are in most cases strongly localized. Through these works it was recognized that the approximate methods cannot provide us with reliable informations concerning the spectral properties of disordered systems, and their nature must be examined critically.

As for the electronic-energy spectrum, the attempt at numerical computation was made somewhat earlier than for the vibrational one. The pioneers were Landauer and Helland (1954), who tried to calculate the band structure of the spectrum of an electron moving in a one-dimensional random array of square-well potentials. Afterwards Lax and Phillips (1958) and Frisch and Lloyd (1960) computed the spectra for a one-dimensional assembly of delta potentials distributed at random. The fine structure was not found, however, until in 1964 Agacy and Borland computed the spectrum for a disordered diatomic Kronig–Penney model. Borland (1963) calculated the wave functions of the electron in a liquid-like array of delta potentials and found that they are, in general, strongly localized.

In the case of electronic systems, the main effort seems to have been rather directed to solving the problem of whether or not the spectral gaps remain when disorder is introduced. As early as 1949 Saxon and Hutner presented a conjecture that in the case of a one-dimensional diatomic mixed chain the interval which corresponds to a spectral gap for both of the pure chains would remain to give a gap for the mixed chain also. This conjecture was proved to be true by Luttinger in 1951. Since then a number of theoretical and numerical studies have been made on the problem of gap persistence, in particular on the problem of whether the theorem of Saxon and Hutner is valid or not for more general systems, but no definite result was obtained until Borland (1961) and Roberts and Makinson (1962) gave an exact condition for gap persistence in the case of a liquid-type array of delta potentials. Roberts and Makinson presented in the same work an ingenious explanation for the localization of wave functions, which was later made more precise by Borland (1963). The argument of these authors is based on a characteristic property of the phase of a wave function which is exhibited when one pursues its spatial variation from one end of the system towards another. For the gap-persistence problem also, the approximate theories failed to yield any reliable conclusion.

A semi-empirical explanation for the peak structure of the spectrum was already given by Dean in the paper mentioned above. He showed numerically that each distinct peak just corresponds to an impurity frequency of an appropriate "island" of light impurities. Hori and Fukushima (1963) gave a theoretical basis for this finding, and showed that the peak structure is just what is to be expected theoretically. They used the method of transfer matrix which had previously been used by Kerner (1954) to discuss the band structure of the energy spectrum of an electron in a random medium and applied by Hori and Asahi (1957) to the problem of vibration of an imperfect chain. Also in the work of Hori and Fukushima the characteristic behaviour of the "phase" played the central role. The formalism of transfer matrix is so general that it can be applied to a wide variety of systems, and the peak structure of the electronic spectrum can be explained in the same way as in the vibrational case.

Again, by using the transfer matrix and the phase, Hori and Matsuda (1964) obtained a very general theorem which states that in a one-dimensional case the statement of the Saxon–Hutner-type is valid for every kind of disordered system which can be described by the transfer matrix. It was shown that not only the results obtained by Luttinger, Borland, and Roberts and Makinson are included in this theorem as special cases, but several numerical results obtained for spectral gaps can be explained by it. Moreover, the theorem can be applied to one and the same system in various different ways, and by doing it appropriately one may obtain quite unexpected conclusions which are completely beyond the scope of the ordinary Saxon–Hutner theorem. Indeed, Hori (1964b), Matsuda and Okada (1965), and Wada (1966) showed that each deep valley appearing in the spectra obtained by Dean, Borland, and others can thus be explained as a narrow spectral gap. The spectral density vanishes at the frequencies corresponding to the bottoms of these valleys. The peculiar nature of these points, called special frequencies or energies, was first noted by Matsuda (1964) and Borland (1964).

The phase, which plays a basic role in the above-mentioned theories, is a generalization of that which appeared in the arguments of Borland and Roberts and Makinson; actually it was already playing an important role implicitly in the theory of Dyson and its variants, and the method of negative factor counting. However, by using the formalism of the transfer matrix one can define the phase with a full generality and construct a unified *phase theory* which includes all the specific treatments hitherto presented as special cases, and can be applied to a wide variety of systems without the necessity of discriminating either vibrational systems from electronic ones or microscopic systems from macroscopic ones. It can be applied to the multi-dimensional systems as well. The purpose of the present book is to give a full account of the phase theory.

In Chapter 1 basic equations of the phase theory and the fundamental notions such as transfer matrix, state ratio, phase, and so on, are introduced. The universal nature of the theory necessarily makes the description somewhat formal, but to avoid excessive abstraction several concrete lattice models, which will be discussed in later chapters, are described, with corresponding specific equations. Phase-theoretical treatment of the spectra of some regular lattices is also given. This is, however, strictly limited to what is needed for showing the use of the theory for regular systems and for reference in later chapters, because the spectra of such systems are well known and there exist several descriptions based upon more usual methods in standard textbooks (see the Bibliographical Notes).

As was mentioned above, many numerical, or model experimental, data have been published with which we can compare the theoretical conclusions. A review of these calculations is given in Chapter 2. In order to inform the reader of the way in which the numerical data were obtained, and also to give a short summary of the interrelation between the phase theory, Dyson's theory and its variants, and the methods of negative factor counting and the Green function, some of the computational methods are outlined in the first part of this chapter.

Chapter 3 is devoted to a description of basic principles of the phase theory with a derivation of the above-mentioned general theorem on the spectral gaps of mixed system.

In Chapter 4 the phase-theoretical treatment of vibrational impurity modes and electronic impurity levels of one-dimensional systems containing few impurities is described. The purpose of this chapter is to show how the phase theory provides us with a description of impurity modes which is far simpler than Green-function theory, on the one hand, and to prepare some notions pertaining to the behaviour of phase which prove to be very useful in later chapters, on the other.

The characteristics of the phase theory are exhibited most conspicuously in Chapter 5, where the general theorems and the notions obtained in Chapters 3 and 4 are applied to several one-dimensional disordered systems, and it is shown that not only the characteristic features of their spectral property, such as peak structure and localization of eigenmodes, but also most of the computational results concerning the gap persistence described in Chapter 2, can be explained by the phase theory from a unified point of view. Especially, the existence of special frequencies and energies is deduced from the general theorem.

The study of one-dimensional models is of value in that many of the inessential complications often associated with higher-dimensional models are absent, so that we can construct a fully perspective theory which provides us with a deep insight into more involved multi-dimensional problems. Moreover, the one-dimensional models are not always unphysical. For example, most of the biopolymers are essentially one-dimensional aperiodic systems, and the study of their vibrational and electronic properties is important in biophysics.

The domain of validity of the phase theory is not restricted to the one-dimensional system, however. For the few-impurity problem the generalization to multi-dimensional case is straightforward. Although the conclusions derived from the phase theory are the same as those drawn from the ordinary Green-function theory, its use for higher-dimensional few-impurity problems is described in detail in Chapter 6 in order to make up for the defect of existing books and review articles which are exlusively based on the Green-function method (see the Bibliographical Notes). It is shown that the phase theory can treat both the point and extended impurities in a unified manner, and yields a simple method of obtaining the equation for the eigenfrequencies of associated impurity modes. The method of transfer matrix, the mother of the phase theory, to which we must often have recourse, and that of scattering matrix, which is a variant of the phase theory and useful for discussing the surface modes, are also described in this chapter.

For disordered multi-dimensional systems, the mathematical difficulty becomes much more pronounced, so that the quantitative conclusions hitherto obtained are limited compared to the abundance of exact theoretical results in the one-dimensional case. A number of qualitative results have already been obtained, however. It was shown that the peak structure of the spectrum may be explained in just the same way as in the one-dimensional case. Matsuda and Okada (1965) succeeded in obtaining a Saxon–Hutner-type theorem by using the continued-fraction theory. From this theorem one can infer that there may exist no special frequency in the spectra of higher-dimensional lattices. It was also shown, however, that special frequencies in a certain generalized sense may still appear in the spectrum. A detailed account of these materials is given in Chapter 6 also.

The aim of this book is to explain the use of the phase theory for few-impurity and disordered-lattice problems, and the account of approximate theories lies outside its scope. However, in order to discuss the reliability of such theories from the standpoint of the phase theory, a review of existing approximate treatments is given in Chapter 7, and their results are compared with the phase-theoretical and numerical ones. This chapter will be of use, though in a limited extent, as a concise summary of the present status of the approximate treatment of disordered lattices.

CHAPTER 1

Basic Equations and Models

1.1. Difference Equations. Vibrational Problems

Consider the system of vector-difference equations of the second order

$$\mathbf{C}_n\mathbf{v}_n + \mathbf{D}_n\mathbf{v}_{n+1} + \mathbf{D}_{n-1}^T\mathbf{v}_{n-1} = 0, \quad n = 1, 2, \ldots, N, \quad (1.1.1)$$

where $\mathbf{v}_n$ is an M-dimensional column vector, $\mathbf{C}_n$ and $\mathbf{D}_n$ are square matrices of order M, and the superscript T means the transpose. $\mathbf{C}_n$ is further assumed to be symmetric and to contain a parameter λ, which is, in fact, a function of energy or frequency of the system and is called the *energy parameter*.

If we impose on (1.1.1) the fixed boundary condition

$$\mathbf{v}_0 = \mathbf{0}, \quad \mathbf{v}_{N+1} = \mathbf{0}, \quad (1.1.2)$$

the condition for (1.1.1) to have a non-trivial solution is

$$|\mathbf{S}(\lambda)| = 0, \quad (1.1.3)$$

where

$$\mathbf{S}(\lambda) = \begin{pmatrix} \mathbf{C}_1 & \mathbf{D}_1 & & & & \\ \mathbf{D}_1^T & \mathbf{C}_2 & \mathbf{D}_2 & & \mathbf{0} & \\ & \mathbf{D}_2^T & \mathbf{C}_3 & \mathbf{D}_3 & & \\ & & & \ddots & & \\ & \mathbf{0} & & \mathbf{D}_{N-2} & \mathbf{C}_{N-1} & \mathbf{D}_{N-1} \\ & & & & \mathbf{D}_{N-1}^T & \mathbf{C}_N \end{pmatrix}. \quad (1.1.4)$$

Equation (1.1.3) is the so-called *secular equation*, each root of which gives an eigenvalue λ_μ and correspondingly an *eigenenergy* or *eigenfrequency*. For each eigenvalue there exists a non-trivial solution of (1.1.1) called an *eigenmode* (an *eigenfunction* or a *normal mode*). The matrix $\mathbf{S}(\lambda)$ and its determinant $|\mathbf{S}(\lambda)|$ are called *secular matrix* and *secular determinant* respectively.

The distribution of λ_μ, or that of the corresponding eigenenergy or eigenfrequency, is the *spectrum* (*energy* or *eigenfrequency spectrum*) of the system. If the system has an infinite extension, that is, if N and/or M are infinite, the spectrum may have continuous parts, where the *spectral density* $D(\mu)$ can be defined, such that $D(\mu)\, d\mu$ gives the fraction of eigenvalues lying between μ and $\mu + d\mu$. The term spectral property means the property of the system which is related directly to the eigenvalues and eigenmodes—property of the spectrum itself, functional form of the normal modes, etc.

Dynamical equations of several atomic lattices can be written in the above form in the approximation of nearest-neighbour harmonic interaction. As an example, consider a two-dimensional square lattice, composed of $N \times M$ particles situated at the integral points $x = n$, $y = m$ in a rectangle $1 \leqq x \leqq N$, $1 \leqq y \leqq M$ in the (x, y)-plane and interacting with one another through nearest-neighbour harmonic coupling (Dean and Martin (1960b)). Its in-plane vibration may be separated into independent motions in x- and y-directions. If a harmonic time-dependence with circular frequency ω is assumed, the equation of motion, e.g. in x-direction, is

$$
\begin{aligned}
-\omega^2 m_{nm} v_{nm} = {} & K'_{nm;\,n,\,m-1}(v_{n,\,m-1} - v_{nm}) + K'_{nm;\,n,\,m+1}(v_{n,\,m+1} - v_{nm}) \\
& + K_{nm;\,n-1,\,m}(v_{n-1,\,m} - v_{nm}) + K_{nm;\,n+1,\,m}(v_{n+1,\,m} - v_{nm}), \\
& n = 1, 2, \ldots, N;\; m = 1, 2, \ldots, M, \qquad (1.1.5)
\end{aligned}
$$

where m_{nm} is the mass of the particle at (n, m), v_{nm} is the x-component of its displacement from the equilibrium position, and $K'_{nm;\,n'm'}$ and $K_{nm;\,n'm'}$ $(= K_{n'm';\,nm})$ are the noncentral and central force constants, respectively, between the particles at (n, m) and (n', m').

Define the matrices

$$\mathbf{M}_n \equiv \operatorname{diag}(m_{n1}, \quad m_{n2}, \ldots, m_{nM}), \tag{1.1.6}$$

$$\mathbf{K}'_n \equiv \left[\begin{array}{c|c|c} \begin{array}{l} K'_{n1,n0} + K'_{n1,n2} \\ + K_{n1;n-1,1} + K_{n1;n+1,1} \end{array} & -K'_{n1,n2} & 0 \\ \hline -K'_{n2,n1} & \begin{array}{l} K'_{n2,n1} + K'_{n2,n3} \\ + K_{n2;n-1,2} + K_{n2;n+1,2} \end{array} & \\ \hline & \ddots & \\ \hline 0 \qquad -K'_{nM;n,M-1} & & \begin{array}{l} K'_{nM;n,M-1} + K'_{nM;n,M+1} \\ + K_{nM;n-1,M} + K_{nM;n+1,M} \end{array} \end{array}\right] \tag{1.1.7}$$

and

$$\mathbf{K}_n \equiv \operatorname{diag}(K_{n1;n+1,1}, \quad K_{n2;n+1,2}, \ldots, K_{nM;n+1,M}), \tag{1.1.8}$$

and put $\omega^2 = \lambda$. Then (1.1.5) can be written

$$(\mathbf{K}'_n - \lambda \mathbf{M}_n)\, \mathbf{v}_n - \mathbf{K}_n \mathbf{v}_{n+1} - \mathbf{K}^T_{n-1} \mathbf{v}_{n-1} = \mathbf{0}, \quad n = 1, 2, \ldots, N, \tag{1.1.9}$$

where $\mathbf{v}_n$ is the vector defined by $\mathbf{v}_n = (v_{n1}, v_{n2}, \ldots, v_{nM})^T$. This equation is just of the form (1.1.1).

The equation for the one-dimensional case is obtained from (1.1.5) by putting the non-central force constants to zero and omitting the suffix m:

$$\omega^2 m_n v_n = (K_{n,n-1} + K_{n,n+1})\, v_n - K_{n,n-1} v_{n-1} - K_{n,n+1} v_{n+1}, \quad n = 1, 2, \ldots, N. \tag{1.1.10}$$

In this case the matrices $\mathbf{M}_n$, $\mathbf{K}'_n$, and $\mathbf{K}_n$ reduce to scalars m_n, $K_{n,n-1} + K_{n,n+1}$ and $K_{n,n+1}$, respectively. When, in particular, the lattice is isotopic so that all the force constants are equal to a constant K, (1.1.10) becomes

$$\omega^2 m_n v_n / K = 2v_n - v_{n-1} - v_{n+1}, \quad n = 1, 2, \ldots, N, \tag{1.1.11}$$

and $\mathbf{M}_n$, $\mathbf{K}'_n$, and $\mathbf{K}_n$ reduce to m_n, $2K$, and K respectively.

As another example, consider a honeycomb lattice as shown in Fig. 1.1 (Dean and Bacon (1965)). Now the motions in two directions are no longer independent. There are two types of lattice site, denoted by α and β in the figure. It can readily be verified that the equations of motion for an atom on α site (n, m) may be written

$$-\omega^2 m_{nm}\begin{pmatrix}u_{nm}\\ v_{nm}\end{pmatrix} = K_{nm;\,n-1,m}\begin{pmatrix}0 & 0\\ 0 & 1\end{pmatrix}\begin{pmatrix}u_{n-1,\,m}\\ v_{n-1,\,m}\end{pmatrix}$$

$$+\begin{bmatrix} -\frac{3}{4}K_{nm;\,n+1,\,m-1} - \frac{3}{4}K_{nm;\,n+1,\,m} & -\frac{\sqrt{3}}{4}K_{nm;\,n+1,\,m-1} + \frac{\sqrt{3}}{4}K_{nm;\,n+1,\,m} \\ -\frac{\sqrt{3}}{4}K_{nm;\,n+1,\,m-1} + \frac{\sqrt{3}}{4}K_{nm;\,n+1,\,m} & -\frac{1}{4}K_{nm;\,n+1,\,m-1} - \frac{1}{4}K_{nm;\,n+1,\,m} - K_{nm;\,n-1,\,m}\end{bmatrix}\begin{bmatrix}u_{nm}\\ v_{nm}\end{bmatrix}$$

$$+ K_{nm;\,n+1,\,m-1}\begin{bmatrix}\frac{3}{4} & \frac{\sqrt{3}}{4}\\ \frac{\sqrt{3}}{4} & \frac{1}{4}\end{bmatrix}\begin{bmatrix}u_{n+1,\,m-1}\\ v_{n+1,\,m-1}\end{bmatrix} + K_{nm;\,n+1,\,m}\begin{bmatrix}\frac{3}{4} & -\frac{\sqrt{3}}{4}\\ -\frac{\sqrt{3}}{4} & \frac{1}{4}\end{bmatrix}\begin{bmatrix}u_{n+1,\,m}\\ v_{n+1,\,m}\end{bmatrix}, \tag{1.1.12}$$

where u_{nm} and v_{nm} represent the components in the directions indicated in Fig. 1.1 of the displacement from equilibrium of the atom at (n, m) and $K_{nm;\,n'm'}$ is the central harmonic force constant between the atoms at (n, m) and (n', m'). For an atom at a β site $(n + 1, m)$, the equation of motion is

$$-\omega^2 m_{n+1,\,m}\begin{bmatrix}u_{n+1,\,m}\\ v_{n+1,\,m}\end{bmatrix} = K_{n+1,\,m;\,nm}\begin{bmatrix}\frac{3}{4} & -\frac{\sqrt{3}}{4}\\ -\frac{\sqrt{3}}{4} & \frac{1}{4}\end{bmatrix}\begin{bmatrix}u_{nm}\\ v_{nm}\end{bmatrix}$$

$$+ K_{n+1,\,m;\,n,\,m+1}\begin{bmatrix}\frac{3}{4} & \frac{\sqrt{3}}{4}\\ \frac{\sqrt{3}}{4} & \frac{1}{4}\end{bmatrix}\begin{bmatrix}u_{n,\,m+1}\\ v_{n,\,m+1}\end{bmatrix}$$

$$+\begin{bmatrix} -\frac{3}{4}K_{n+1,\,m;\,nm} - \frac{3}{4}K_{n+1,\,m;\,n,\,m+1} & \frac{\sqrt{3}}{4}K_{n+1,\,m;\,nm} - \frac{\sqrt{3}}{4}K_{n+1,\,m;\,n,\,m+1} \\ \frac{\sqrt{3}}{4}K_{n+1,\,m;\,nm} - \frac{\sqrt{3}}{4}K_{n+1,\,m;\,n,\,m+1} & -\frac{1}{4}K_{n+1,\,m;\,nm} - \frac{1}{4}K_{n+1,\,m;\,n,\,m+1} - K_{n+1,\,m;\,n+2,\,m}\end{bmatrix}\begin{bmatrix}u_{n+1,\,m}\\ v_{n+1,\,m}\end{bmatrix}$$

$$+ K_{n+1,\,m;\,n+2,\,m}\begin{pmatrix}0 & 0\\ 0 & 1\end{pmatrix}\begin{pmatrix}u_{n+2,\,m}\\ v_{n+2,\,m}\end{pmatrix}. \tag{1.1.13}$$

If we define the vector

$$\mathbf{v}_n \equiv (u_{n1}, v_{n1}, u_{n2}, v_{n2}, \ldots, u_{nM}, v_{nM})^T$$

and the matrices

$$\mathbf{M}_n = \operatorname{diag}(m_{n1}, m_{n1}, m_{n2}, m_{n2}, \ldots, m_{nM}, m_{nM}), \tag{1.1.14}$$

$$\mathbf{K}_n'^{\alpha} \equiv \frac{1}{4} \left[\begin{array}{c|c} \ddots & \\ 3\begin{pmatrix} K_{nm;\,n+1,\,m-1} \\ \quad + K_{nm;\,n+1,\,m} \end{pmatrix} & 3\begin{pmatrix} K_{nm;\,n+1,\,m-1} \\ \quad - K_{nm;\,n+1,\,m} \end{pmatrix} \\ \hline \sqrt{3}\begin{pmatrix} K_{nm;\,n+1,\,m-1} \\ \quad - K_{nm;\,n+1,\,m} \end{pmatrix} & \begin{pmatrix} K_{nm;\,n+1,\,m-1} + K_{nm;\,n+1,\,m} \\ + 4K_{nm;\,n-1,\,m} \end{pmatrix} \\ & \ddots \end{array} \right], \tag{1.1.15}$$

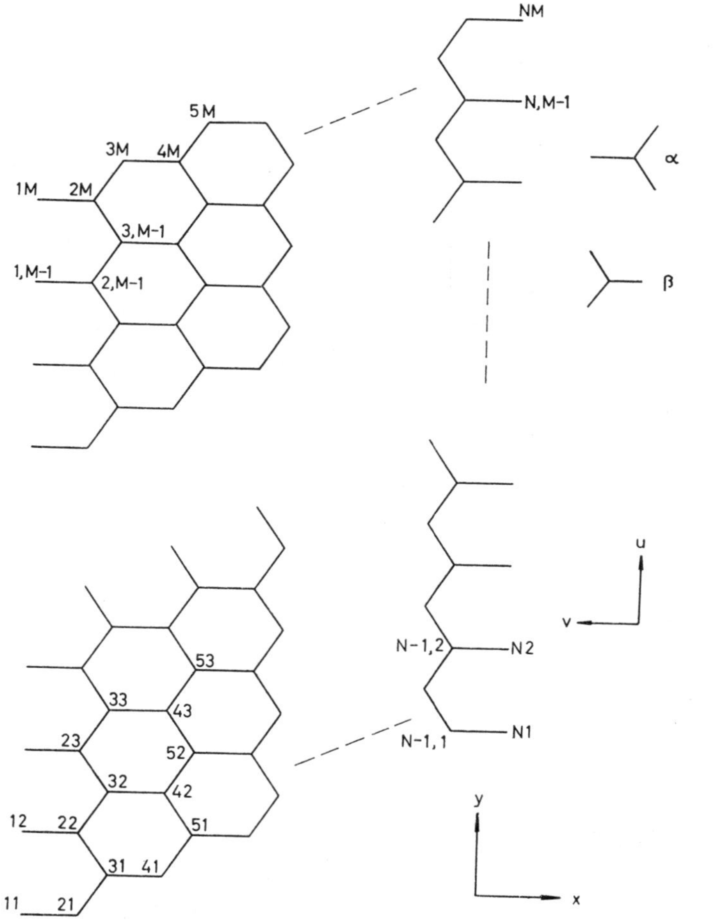

FIG. 1.1. Enumeration of lattice sites for the honeycomb lattice.

$$\mathbf{K}_n'^{\beta} \equiv \tfrac{1}{4}\left[\begin{array}{c|c} \ddots & \\ 3\begin{pmatrix} K_{nm;\,n-1,\,m} \\ \quad + K_{nm;\,n-1,\,m+1}\end{pmatrix} & \sqrt{3}\begin{pmatrix} K_{nm;\,n-1,\,m+1} \\ \quad - K_{nm;\,n-1,\,m}\end{pmatrix} \\ \hline \sqrt{3}\begin{pmatrix} K_{nm;\,n-1,\,m+1} \\ \quad - K_{nm;\,n-1,\,m}\end{pmatrix} & \begin{pmatrix} K_{nm;\,n-1,\,m} + K_{nm;\,n-1,\,m+1} \\ \quad + 4K_{nm;\,n+1,\,m}\end{pmatrix} \\ & \ddots \end{array}\right], \tag{1.1.16}$$

$$\mathbf{K}_n^{\alpha} \equiv$$

$$\left[\begin{array}{c|c|c} \ddots & K_{nm;\,n+1,\,m-1}\begin{pmatrix} \frac{3}{4} & \frac{\sqrt{3}}{4} \\ \frac{\sqrt{3}}{4} & \frac{1}{4}\end{pmatrix} & \\ \hline K_{nm;\,n+1,\,m-1}\begin{pmatrix} \frac{3}{4} & \frac{\sqrt{3}}{4} \\ \frac{\sqrt{3}}{4} & \frac{1}{4}\end{pmatrix} & K_{nm;\,n+1,\,m}\begin{pmatrix} \frac{3}{4} & -\frac{\sqrt{3}}{4} \\ -\frac{\sqrt{3}}{4} & \frac{1}{4}\end{pmatrix} & K_{n,\,m+1;\,n+1,\,m}\begin{pmatrix} \frac{3}{4} & \frac{\sqrt{3}}{4} \\ \frac{\sqrt{3}}{4} & \frac{1}{4}\end{pmatrix} \\ \hline & K_{n,\,m+1;\,n+1,\,m}\begin{pmatrix} \frac{3}{4} & \frac{\sqrt{3}}{4} \\ \frac{\sqrt{3}}{4} & \frac{1}{4}\end{pmatrix} & \ddots \end{array}\right], \tag{1.1.17}$$

$$\mathbf{K}_n^{\beta} \equiv \begin{pmatrix} \ddots & & \\ & K_{n+1,\,m;\,nm}\begin{pmatrix} 0 & 0 \\ 0 & 1\end{pmatrix} & \\ & & \ddots \end{pmatrix}, \tag{1.1.18}$$

and put $\lambda = \omega^2$, the equations of motion can be written in the form:

$$\lambda \mathbf{M}_n \mathbf{v}_n = \mathbf{K}_n'^{\alpha}\mathbf{v}_n - \mathbf{K}_n^{\alpha}\mathbf{v}_{n+1} - \mathbf{K}_{n-1}^{\beta^T}\mathbf{v}_{n-1}, \quad \text{for } \alpha \text{ sites}, \tag{1.1.19a}$$

$$\lambda \mathbf{M}_n \mathbf{v}_n = \mathbf{K}_n'^{\beta}\mathbf{v}_n - \mathbf{K}_n^{\beta}\mathbf{v}_{n+1} - \mathbf{K}_{n-1}^{\alpha^T}\mathbf{v}_{n-1}, \quad \text{for } \beta \text{ sites}. \tag{1.1.19b}$$

This is again of the form (1.1.1).

It is sometimes convenient to let the matrix $\mathbf{C}_n$ have the form $\mathbf{A}_n - \lambda\mathbf{I}$, where $\mathbf{A}_n$ no longer contains the energy parameter, and $\mathbf{I}$ is the unit matrix of order M. For this purpose we have only to multiply (1.1.9), for example, from the left by $\mathbf{M}_n^{-1/2}$ and put $\mathbf{w}_n = \mathbf{M}_n^{1/2}\mathbf{v}_n$. Then it becomes

$$(\mathbf{A}_n - \lambda\mathbf{I})\,\mathbf{w}_n - \mathbf{B}_n\mathbf{w}_{n+1} - \mathbf{B}_{n-1}^T\mathbf{w}_{n-1}, \tag{1.1.20}$$

where we have defined

$$\mathbf{A}_n \equiv \mathbf{M}_n^{-1/2}\mathbf{K}_n'\mathbf{M}_n^{-1/2}, \tag{1.1.21}$$

and

$$\mathbf{B}_n \equiv \mathbf{M}_n^{-1/2}\mathbf{K}_n\mathbf{M}_{n+1}^{-1/2}. \tag{1.1.22}$$

The secular matrix of (1.1.20) is

$$\mathbf{S}(\lambda) \equiv \mathbf{P} - \lambda\mathbf{I}$$

$$= \begin{pmatrix} \mathbf{A}_1 - \lambda\mathbf{I} & -\mathbf{B}_1 & & & & \\ -\mathbf{B}_1^T & \mathbf{A}_2 - \lambda\mathbf{I} & -\mathbf{B}_2 & & & \\ & -\mathbf{B}_2^T & \mathbf{A}_3 - \lambda\mathbf{I} & \ddots & & \\ & & & \ddots & & \\ & & & & \mathbf{A}_{N-1} - \lambda\mathbf{I} & -\mathbf{B}_{N-1} \\ & & & & -\mathbf{B}_{N-1}^T & \mathbf{A}_N - \lambda\mathbf{I} \end{pmatrix}. \tag{1.1.23}$$

In this case the eigenvalues λ_μ are, by definition, those of the matrix $\mathbf{P}$ which is independent of λ and is called *dynamical matrix*.

For the first of the above examples, the matrices $\mathbf{A}_n$ and $\mathbf{B}_n$ are given by

$$\mathbf{A}_n \equiv \begin{pmatrix} a_{n1} & b_{n1} & & & & \\ b_{n1} & a_{n2} & b_{n2} & & & \\ & b_{n2} & a_{n3} & \ddots & & \\ & & & \ddots & & \\ & & & & a_{n,M-1} & b_{n,M-1} \\ & & & & b_{n,M-1} & a_{nM} \end{pmatrix}, \tag{1.1.24}$$

and

$$\mathbf{B}_n \equiv \operatorname{diag}(c_{n1}, c_{n2}, \ldots, c_{nM}), \tag{1.1.25}$$

respectively, where

$$a_{nm} \equiv (K'_{nm;n,m-1} + K'_{nm;n,m+1} + K_{nm;n-1,m} + K_{nm;n+1,m})/m_{nm}, \tag{1.1.26}$$

$$b_{nm} \equiv -K'_{n,m+1;nm}/(m_{n,m+1}m_{nm})^{1/2}, \tag{1.1.27}$$

and

$$c_{nm} \equiv K_{nm;n+1,m}/(m_{nm}m_{n+1,m})^{1/2}. \tag{1.1.28}$$

In the one-dimensional case, the matrices $\mathbf{A}_n$ and $\mathbf{B}_n$ are reduced to scalars

$$A_n = (K_{n,n-1} + K_{n,n+1})/m_n \quad \text{and} \quad B_n = K_{n,n+1}/(m_n m_{n+1})^{1/2} \tag{1.1.29}$$

respectively.

The above formalism can, of course, be generalized to three-dimensional lattices. The generalization is utterly straightforward, however, leading to a mere repetition of similar formulae.

It is to be noted that in the above we considered the fixed boundary condition only. To introduce other boundary conditions merely amounts to slightly modifying the matrix elements pertaining to the boundaries. Such modifications are discussed in § 1.5.

1.2. Electronic Systems Described by Difference Equations

The difference equation of the type (1.1.1) can describe some electronic systems as well. For example, consider a polymer consisting of a random sequence in x-direction of two kinds of monomer which are not too much different from each other (Herzenberg and Modinos (1965)). Suppose that each monomer interacts with its nearest neighbours through an electrostatic dipole–dipole potential and that the exchange of electrons can be ignored. Then the states of the complete system are given by the sums of products of the wave function $\psi_{n\alpha}$ which describes the state α of the nth molecule ($\alpha = 0, 1, 2, \ldots; n = 1, 2, \ldots, N$). In the absence of interactions, the ground state is

$$\Psi_0 = \prod_{n=1}^{N} \psi_{n0}, \tag{1.2.1}$$

and the simplest excited states are those in which just a single monomer is excited, namely

$$\Psi_{n\alpha} = \Psi_0(\psi_{n\alpha}/\psi_{n0}). \tag{1.2.2}$$

Denote the energy of the nth molecule in the state α by $E_{n\alpha}$. By the assumption of small difference between the monomers, the difference between $E_{n\alpha}$'s for a fixed α is small compared to the spacing of $E_{n\alpha}$ of a single monomer. Then one may put

$$\Phi_\alpha = \sum_{n=1}^{N} v_n \Psi_{n\alpha}, \tag{1.2.3}$$

so that the expectation value of the energy is

$$E = \int \Phi_\alpha H \Phi_\alpha \, dv \Big/ \int \Phi_\alpha^2 \, dv = \Big(v_n^2 H_{nn} + \sum_{n'=n\pm 1} v_n v_{n'} H_{nn'} \Big) \Big/ \sum_n v_n^2, \tag{1.2.4}$$

where $H \equiv H_0 + V$ is the total Hamiltonian, H_0 and V being the unperturbed part and the dipole–dipole interaction respectively,

and

$$H_{nn'} = (\Psi_{n\alpha}|H|\Psi_{n'\alpha}). \tag{1.2.5}$$

The overlap between $\Psi_{n\alpha}$'s has been neglected in (1.2.4).

The condition for E to be minimum is obtained from the equation

$$\partial E/\partial v_n = 0, \quad n = 1, 2, ..., N. \tag{1.2.6}$$

From (1.2.4) and (1.2.6) we get

$$(E_{n\alpha} - E)\, v_n + \sum_{n'=n\pm 1} (\Psi_{n\alpha}|V|\Psi_{n'\alpha})\, v_{n'} = 0, \tag{1.2.7}$$

since we may put

$$(\Psi_{n\alpha}|H|\Psi_{n\alpha}) = E_{n\alpha}. \tag{1.2.8}$$

The matrix elements of V can be written

$$(\Psi_{n\alpha}|V|\Psi_{n'\alpha}) = \frac{1}{R_{nn'}^3}\left[\frac{3\mathbf{P}_{n\alpha 0}\cdot \boldsymbol{R}_{nn'}\mathbf{P}_{n'0\alpha}\cdot \boldsymbol{R}_{nn'}}{R_{nn'}^2} - \mathbf{P}_{n\alpha 0}\cdot \mathbf{P}_{n'0\alpha}\right]. \tag{1.2.9}$$

Here $\mathbf{P}_{n\alpha 0} \equiv (\psi_{n\alpha}|\mathbf{P}_n|\psi_{n0})$, $\mathbf{P}_n$ being the dipole operator of the nth monomer. $\boldsymbol{R}_{nn'}$ is the vector from monomer n to monomer n', and $R_{nn'} \equiv |\boldsymbol{R}_{nn'}|$.

Let us assume that each monomer has three orthogonal planes of mirror symmetry, one of which is perpendicular to the x-direction. Then the electronic states may be classified by the triplets of the reflection eigenvalues (± 1, ± 1, ± 1), where any combination of plus and minus signs is possible. The ground state is assumed to be ($+1$, $+1$, $+1$). It is sufficient to consider only those excited states Φ whose reflection eigenvalues consist of two ($+1$)'s and one (-1), since only such states have a non-vanishing dipole transition moment to the ground state. The matrix elements $\mathbf{P}_{n\alpha 0}$ and $\mathbf{P}_{n0\alpha}$ have a single non-vanishing component in the direction corresponding to the eigenvalue (-1). These components are denoted by $P_{n\alpha 0}$ and $P_{n0\alpha}$ respectively. Thus (1.2.9) becomes

$$(\Psi_{n\alpha}|V|\Psi_{n'\alpha}) = gP_{n\alpha 0}P_{n'0\alpha}/R_{nn'}^3 \equiv P_{nn'}, \tag{1.2.10}$$

where $g = +2$ for transition moments parallel to x-direction and -1 for the normal direction. Therefore (1.2.7) reduces to

$$(E_{n\alpha} - E)\, v_n + P_{n,n+1}v_{n+1} + P_{n,n-1}v_{n-1} = 0, \tag{1.2.11}$$

which is again a difference equation of the type (1.1.1), where E plays the role of energy parameter.

Next consider, as another example, a one-dimensional array of N identical potentials $V(x - x_n)$, whose centres lie at x_n. Let $\psi(x)$

be the wave function (atomic orbital) for a single electron moving in an isolated potential. Then the zeroth-order wave function for the array is given by $\psi(x - x_n)$, in the *tight-binding approximation.* There are N such wave functions, and the electron can be bound to any one of the N possible sites. A better approximation may be obtained by constructing an orthogonal set $\{\Psi_n\}$ using the Schmidt orthogonalization process. If we assume that

$$S_{n'n} \equiv \int_{-\infty}^{+\infty} \psi^*(x - x_{n'})\, \psi(x - x_n)\, dx \tag{1.2.12}$$

is negligible unless $n' = n$ or $n' = n \pm 1$, Ψ_n is given by, to the first order in $S_{n'n}$,

$$\Psi_n = -\frac{S_{n,\,n-1}}{(1 - S_{n,\,n-1}^2)^{1/2}}\, \psi(x - x_{n-1}) + \frac{1}{(1 - S_{n,\,n-1}^2)^{1/2}}\, \psi(x - x_n). \tag{1.2.13}$$

We now use the wave functions Ψ_n to calculate the matrix elements of the one-electron Hamiltonian

$$H = -d^2/dx^2 + \sum_{n=1}^{N} V(x - x_n), \tag{1.2.14}$$

where we have put $\hbar^2/2m = 1$. (We shall use this simplification throughout this book.) If we neglect Coulomb interactions between electrons, the energy levels of the system of N electrons moving in the array are given by eigenvalues of this Hamiltonian. Define the quantity $W_{n'n}$ by

$$W_{n'n} \equiv \int_{-\infty}^{+\infty} \psi^*(x - x_{n'})\, V_n' \psi(x - x_n)\, dx, \tag{1.2.15}$$

where

$$V_n' = \sum_{n' \neq n} V(x - x_{n'}), \tag{1.2.16}$$

and assume that $W_{n'n}$ vanishes unless $n' = n$ or $n \pm 1$. Neglecting squares and higher powers of $W_{n'n}$ and $S_{n'n}$, we find that the only non-zero matrix elements are

$$H_{nn} \equiv \int \Psi_n^* H \Psi_n\, dx = E_1 + (1 - S_{n,\,n-1}^2)^{-1/2}(W_{nn} + 2S_{n,\,n-1} W_{n,n-1}) \tag{1.2.17}$$

and

$$H_{n,\,n+1} \equiv \int \Psi_n^* H \Psi_{n+1} dx = [(1 + S_{n,\,n-1}^2)\,(1 - S_{n+1,\,n}^2)]^{-1/2} \times (W_{n,\,n+1} - S_{n,\,n+1} W_{nn}). \tag{1.2.18}$$

Here E_1 is the ionization energy of an electron in an isolated potential. Thus the dynamical matrix of the present system has the tridiagonal form such as (1.1.4), so that the matrix with the elements $H_{n'n} - E\delta_{n'n}$ can be considered as the secular matrix associated with a difference equation of the form (1.1.1), with the energy parameter E.

The so-called *Hückel approximation* in the theory of electronic states of organic molecules is of similar nature and also leads to a difference equation of the type (1.1.1).

1.3. Transfer Matrix

Let $\mathbf{V}_n$ be the square matrix of order M, the mth column of which is the solution of (1.1.1) satisfying the initial conditions

$$\mathbf{v}_0 = \mathbf{0}, \quad \mathbf{v}_1 = (\underbrace{0, 0, \ldots, 1}_{m}, 0, \ldots, 0)^T. \tag{1.3.1}$$

The matrix $\mathbf{V}_n$ satisfies the matrix-difference equation

$$\mathbf{C}_n\mathbf{V}_n + \mathbf{D}_n\mathbf{V}_{n+1} + \mathbf{D}_{n-1}^T\mathbf{V}_{n-1} = \mathbf{0}, \quad n = 1, 2, \ldots, N, \tag{1.3.2}$$

with the initial condition

$$\mathbf{V}_0 = \mathbf{0}, \quad \mathbf{V}_1 = \mathbf{I}. \tag{1.3.3}$$

The solution of (1.1.1) for an arbitrary initial vector $\mathbf{v}_1$ is given by $\mathbf{V}_n\mathbf{v}_1$.

Equation (1.3.2) can be written in the form of a vector-matrix equation:

$$\begin{pmatrix} \mathbf{V}_n \\ \mathbf{V}_{n+1} \end{pmatrix} = \begin{pmatrix} \mathbf{0} & \mathbf{I} \\ -\mathbf{D}_n^{-1}\mathbf{D}_{n-1}^T & -\mathbf{D}_n^{-1}\mathbf{C}_n \end{pmatrix} \begin{pmatrix} \mathbf{V}_{n-1} \\ \mathbf{V}_n \end{pmatrix}, \tag{1.3.4}$$

where $\mathbf{D}_n$ has been assumed to be non-singular. This equation can be interpreted in words that the "state" of the system at $x = n$ represented by the vector $\mathbf{X} \equiv (\mathbf{V}_{n-1}, \mathbf{V}_n)^T$ is transferred to the next state represented by $\mathbf{X}_{n+1} \equiv (\mathbf{V}_n, \mathbf{V}_{n+1})^T$ through the matrix

$$\mathbf{T}_n \equiv \begin{pmatrix} \mathbf{0} & \mathbf{I} \\ -\mathbf{D}_n^{-1}\mathbf{D}_{n-1}^T & -\mathbf{D}_n^{-1}\mathbf{C}_n \end{pmatrix}. \tag{1.3.5}$$

In this sense, the vector $\mathbf{X}_n$ is called *state vector*, and the matrix $\mathbf{T}_n$ is called *transfer matrix*.

For the case of (1.1.9), the transfer matrix is given by

$$\mathbf{T}_n = \begin{pmatrix} \mathbf{0} & \mathbf{I} \\ -\mathbf{K}_n^{-1}\mathbf{K}_{n-1}^T & \mathbf{K}_n^{-1}(\mathbf{K}_n' - \lambda\mathbf{M}_n) \end{pmatrix}, \tag{1.3.6}$$

which reduces in the one-dimensional case to

$$\mathbf{T}_n = \begin{pmatrix} 0 & 1 \\ -K_{n,n-1}/K_{n,n+1} & (K_{n,n-1}+K_{n,n+1}-\lambda m_n)/K_{n,n+1} \end{pmatrix}. \quad (1.3.7)$$

In the case of an isotopic chain, this is further reduced to

$$\mathbf{T}_n = \begin{pmatrix} 0 & 1 \\ -1 & 2-\lambda m_n/K \end{pmatrix}. \quad (1.3.8)$$

In some cases it is convenient to take the vector $\mathbf{X}_n = (v_{n-1}, v_n - v_{n-1})^T$ as the state vector, instead of $(v_{n-1}, v_n)^T$. The transformation to the new representation is effected by the matrix

$$\mathbf{Y} = \begin{pmatrix} 1 & 0 \\ -1 & 1 \end{pmatrix}, \quad (1.3.9)$$

and the transfer matrix (1.3.7) becomes

$$\mathbf{T}_n = \begin{pmatrix} 1 & 1 \\ -\lambda m_n/K_{n,n+1} & (K_{n,n-1}-\lambda m_n)/K_{n,n+1} \end{pmatrix}, \quad (1.3.10)$$

which reduces to

$$\mathbf{T}_n = \begin{pmatrix} 1 & 1 \\ -m_n\lambda/K & 1-m_n\lambda/K \end{pmatrix}, \quad (1.3.11)$$

in the isotopic case.

For the introduction of the transfer matrix it was necessary to assume the non-singularity of $\mathbf{D}_n$. This is not a severe restriction,

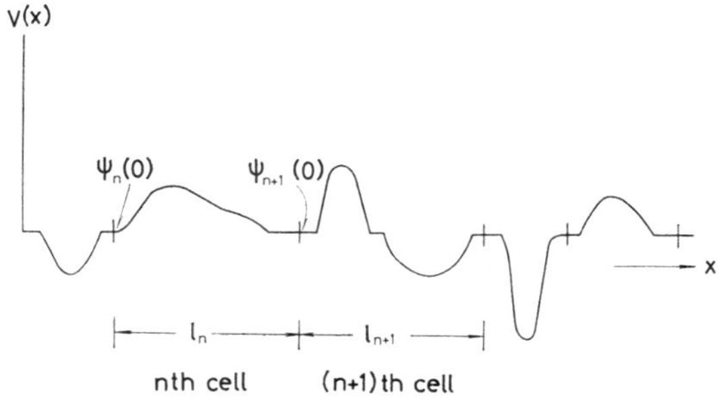

FIG. 1.2. An example of a one-dimensional array of potentials.

however, since $\mathbf{D}_n$ is singular only in few cases such as the second example of § 1.1. The merit of the transfer matrix formalism largely compensates such a minor defect. In the first place, its use is not limited to the systems described by the difference equation. As a typical example, consider an electron moving in a random array of potentials such as shown in Fig. 1.2, where it is assumed that between neighbouring potentials there exists a zero-potential region. Let us divide the system into N cells with the ends in zero-potential regions. Each cell may contain an arbitrary number of potentials, as in the $(n+1)$th cell in the figure. It will be seen later that this arbitrariness plays a very important role in the phase theory.

In each cell, there exist two independent real solutions $y_n^{(1)}(E, x)$ and $y_n^{(2)}(E, x)$ of the Schrödinger equation

$$\{d^2/dx^2 + E - V(x)\}\psi = 0, \tag{1.3.12}$$

which satisfy the boundary conditions

$$y_n^{(1)}(E, 0) = 1, \quad y_n^{(1)\prime}(E, 0) = 0 \tag{1.3.13}$$

and

$$y_n^{(2)}(E, 0) = 0, \quad y_n^{(2)\prime}(E, 0) = 1 \tag{1.3.14}$$

respectively. In each cell the left-hand end of the cell has been chosen as the origin of coordinates in that cell. The general solution of (1.3.12) in the nth cell may be written in terms of these solutions:

$$\psi_n(E, x) = a_n y_n^{(1)}(E, x) + b_n y_n^{(2)}(E, x). \tag{1.3.15}$$

The condition of continuity of ψ and its derivative at the right-hand end of each cell leads to the relation

$$\begin{aligned} \psi_{n+1}(E, 0) &= a_{n+1} = t_n^{(11)} a_n + t_n^{(12)} b_n, \\ \psi'_{n+1}(E, 0) &= b_{n+1} = t_n^{(21)} a_n + t_n^{(22)} b_n, \end{aligned} \tag{1.3.16}$$

where

$$\begin{aligned} t_n^{(11)}(E) &\equiv y_n^{(1)}(E, l_n), \quad t_n^{(12)}(E) \equiv y_n^{(2)}(E, l_n), \\ t_n^{(21)}(E) &\equiv y_n^{(1)\prime}(E, l_n) \quad t_n^{(22)}(E) \equiv y_n^{(2)\prime}(E, l_n). \end{aligned} \tag{1.3.17}$$

Equation (1.3.16) can be written in a vector–matrix form:

$$\begin{pmatrix} \psi_{n+1}(0) \\ \psi'_{n+1}(0) \end{pmatrix} = \begin{pmatrix} t_n^{(11)} & t_n^{(12)} \\ t_n^{(21)} & t_n^{(22)} \end{pmatrix} \begin{pmatrix} \psi_n(0) \\ \psi'_n(0) \end{pmatrix}. \tag{1.3.18}$$

Although this equation cannot be reduced to a difference equation, it has the same form as (1.3.4), and can be interpreted likewise.

For later use, consider the case in which every potential is a delta-type one and each cell contains only one δ-potential at its right-hand end, as shown in Fig. 1.3 (Kronig–Penney model). Let the strength of the δ-potential in the nth cell be $-V_n$. Then it can

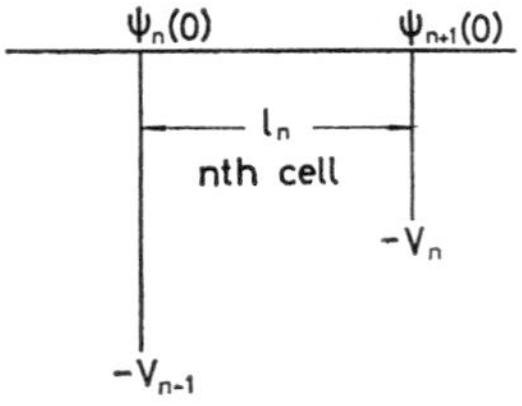

FIG. 1.3. Enumeration of cells and potentials in the disordered Kronig–Penney model.

easily be shown that

$$y_n^{(1)}(E, l_n) = \cos kl_n, \qquad y_n^{(1)\prime}(E, l_n) = -V_n \cos kl_n - k \sin kl_n,$$
$$y_n^{(2)}(E, l_n) = \sin kl_n/k, \quad y_n^{(2)\prime}(E, l_n) = \cos kl_n - (V_n/k) \sin kl_n, \tag{1.3.19}$$

where $k \equiv E^{1/2}$. The transfer-matrix $\mathbf{T}_n$ becomes

$$\mathbf{T}_n = \begin{pmatrix} \cos kl_n & \sin kl_n/k \\ -k \sin kl_n - V_n \cos kl_n & \cos kl_n - (V_n/k) \sin kl_n \end{pmatrix}. \tag{1.3.20}$$

In this case the quantity k plays the role of energy parameter.

The second and the most outstanding merit of the transfer matrix formalism is that it enables one to describe the successive spatial change (in n- or x-direction) of the state of the system as a process of successive transfer by the transfer matrix, which is mathematically described by the matrix multiplication

$$\mathbf{X}_{n+k} = \mathbf{T}_{n+k-1}\mathbf{T}_{n+k-2} \ldots \mathbf{T}_n\mathbf{X}_n. \tag{1.3.21}$$

Moreover, we can regard a product of an arbitrary number of successive matrices as a single transfer matrix. For example, (1.3.21) can be written either

$$\mathbf{X}_{n+k} = \mathbf{T}_{n+k-1;\, n+k-l-1}\mathbf{T}_{n+k-l;\, n}\mathbf{X}_n \tag{1.3.22}$$

or

$$\mathbf{X}_{n+k} = \mathbf{T}_{n+k-1;\, n}\mathbf{X}_n \tag{1.3.23}$$

or in several other ways, where

$$\mathbf{T}_{n;\, l} \equiv \mathbf{T}_n\mathbf{T}_{n-1} \ldots \mathbf{T}_{l+1}\mathbf{T}_l. \tag{1.3.24}$$

This freedom of choice of the set of transfer matrices just corresponds to the above-mentioned arbitrariness in choosing the cells.

It will be seen in the following that the phase theory works well for every system which can be described by the transfer matrix, provided that it is originally governed either by a vector-difference equation of the second order or by a differential equation of Sturm–Liouville type. Therefore the domain of applicability of the phase theory is not confined to molecular systems such as mentioned above. It is applicable also to the systems composed of macroscopic units, such as electric networks and optical multilayer filters. We shall denote the state vector by $\mathbf{X}_n$ and the transfer matrix by $\mathbf{T}_n$ when it is real, irrespective of the physical nature of the system under consideration.

It is to be noted that the determinant of the transfer matrix is usually positive. In what follows we assume that this is always the case. Moreover, in most cases it is unity, that is, the matrix is *unimodular*. For example, the determinant of the transfer matrix in (1.3.18) is just the Wronskian of $y_n^{(1)}$ and $y_n^{(2)}$ at $x = l_n$. The value of the Wronskian is, however, known to be independent of x, and it is unity at $x = 0$ according to (1.3.13).

1.4. Phase Matrix, Linear Transformation, and Continued Fraction

Let us define the matrix $\mathbf{U}_n = \mathbf{X}_n^{(1)}\mathbf{X}_n^{(2)-1}$, where $\mathbf{X}_n^{(1)}$ and $\mathbf{X}_n^{(2)}$ are the components of the state vector $\mathbf{X}_n$, and call it *state ratio matrix* or simply *state ratio*. The relation between the successive state ratio matrices $\mathbf{U}_n$ and $\mathbf{U}_{n+1}$ is easily seen to be

$$\mathbf{U}_{n+1} = [\mathbf{T}_n^{(11)}\mathbf{U}_n + \mathbf{T}_n^{(12)}]\,[\mathbf{T}_n^{(21)}\mathbf{U}_n + \mathbf{T}_n^{(22)}]^{-1}, \qquad (1.4.1)$$

where $\mathbf{T}_n^{(ij)}$ is the (i, j)-element of $\mathbf{T}_n$. In the one-dimensional case, in which the matrices $\mathbf{U}_n$ and $\mathbf{T}_n^{(ij)}$ are reduced to the scalars u_n and $t_n^{(ij)}$ respectively, the relation (1.4.1) becomes

$$u_{n+1} = (t_n^{(11)}u_n + t_n^{(12)})/(t_n^{(21)}u_n + t_n^{(22)}). \qquad (1.4.2)$$

Such a transformation from u_n to u_{n+1} is just a *linear transformation* or *Möbius transformation* as it is called in the theory of functions of complex variable.

In the case of the difference equation, the state ratio $\mathbf{U}_n$ is $\mathbf{V}_{n-1}\mathbf{V}_n^{-1}$, and the linear transformation (1.4.1) becomes a recurrence equation

$$-\mathbf{D}_n\mathbf{U}_{n+1}^{-1} = \mathbf{C}_n + \mathbf{D}_{n-1}^T\mathbf{U}_n, \qquad \mathbf{U}_1 = \mathbf{0} \qquad (1.4.3)$$

or

$$\mathbf{K}_n\mathbf{U}_{n+1}^{-1} = \mathbf{K}_n' - \lambda\mathbf{M}_n - \mathbf{K}_{n-1}^T\mathbf{U}_n, \quad \mathbf{U}_1 = \mathbf{0} \qquad (1.4.4)$$

according as the transfer matrix is given by (1.3.5) or (1.3.6). For the one-dimensional isotopic lattice, (1.4.4) is further reduced to a very simple relation:

$$1/u_{n+1} = 2 - m_n\lambda/K - u_n. \tag{1.4.5}$$

This leads to a *continued-fraction expansion*

$$u_{n+1} = \cfrac{1}{2 - m_n\lambda/K - \cfrac{1}{2 - m_{n-1}\lambda/K - \cfrac{1}{\ddots}}} \tag{1.4.6}$$

Such an expansion can of course be derived from the general equation (1.4.1) as well. The discussion of the general case is postponed to Chapter 3, where the principles of the method of continued fraction is explained.

It should be noted that the definition of the state ratio matrix involves a difficulty, that is, it diverges whenever $\mathbf{X}_n^{(2)}$ is singular. This difficulty may, however, be avoided by considering instead of $\mathbf{U}_n$ itself its *Cayley transform*

$$\mathbf{Z}_n = -(\mathbf{U}_n - i\mathbf{I})(\mathbf{U}_n + i\mathbf{I})^{-1}, \tag{1.4.7}$$

which is always non-singular, in so far as $\mathbf{U}_n$ is real. If $\mathbf{U}_n$ is also symmetric, $\mathbf{Z}_n$ becomes unitary. In such a case the Cayley transform defines a mapping of the space of real symmetric matrices into the space of unitary ones, and one can further introduce the *phase matrix* $\mathbf{\Delta}_n$ of $\mathbf{Z}_n$ defined by

$$\exp(i\mathbf{\Delta}_n) \equiv \mathbf{Z}_n \tag{1.4.8}$$

or

$$\mathbf{\Delta}_n \equiv 2\tan^{-1}\mathbf{U}_n. \tag{1.4.9}$$

To investigate the successive transfer of the state vector amounts to examining the successive change of the phase matrix $\mathbf{\Delta}_n$. In later chapters it will be shown that the quantities $\mathbf{U}_n$, $\mathbf{Z}_n$, and $\mathbf{\Delta}_n$ are very convenient quantities in the sense that, from the investigation of their behaviour, several valuable informations are obtained concerning the spectral properties of the system described by the transfer-matrix formalism. There we shall come across the cases in which both the state vector and transfer matrix, and consequently also the state ratio matrix, are complex from the outset. In such cases the state ratio will be denoted by $\mathbf{Z}_n$ from the beginning.

In the case of one dimension, the matrices $\mathbf{U}_n$, $\mathbf{Z}_n$, and $\mathbf{\Delta}_n$ are reduced to the scalars u_n, z_n and δ_n respectively. The Cayley transform

$$z_n = -(u_n - i)/(u_n + i) \tag{1.4.10}$$

defines a mapping of the real axis on to the unit circle of the complex plane, and δ_n is just the phase angle of the complex number z_n on the circle. The quantities u_n or z_n and δ_n are called simply *state ratio* and *phase* respectively.

The transform (1.4.10) is by no means the unique one which maps the real axis on to the unit circle. Consider a linear transformation

$$z_n = (\alpha u_n + \beta)/(\gamma u_n + \delta), \tag{1.4.11}$$

where α, β, γ, and δ are complex numbers satisfying either the condition

$$\begin{aligned} \alpha\delta &= (\beta\gamma)^*, \quad \alpha\gamma = (\alpha\gamma)^*, \quad \alpha^2 = \gamma^{2*}, \\ \alpha\beta &= (\gamma\delta)^*, \quad \beta\delta = (\beta\delta)^*, \quad \delta^2 = \beta^{2*}, \end{aligned} \tag{1.4.12}$$

or

$$\begin{aligned} \alpha\delta &= (-\beta\gamma)^*, \quad \alpha\gamma = -(\alpha\gamma)^*, \quad \alpha^2 = -\gamma^{2*}, \\ \alpha\beta &= (-\gamma\delta)^*, \quad \beta\delta = -(\beta\delta)^*, \quad \beta^2 = -\delta^{2*}. \end{aligned} \tag{1.4.13}$$

It is easily proved that in both cases $|z_n|^2 = 1$ for any real u_n. Thus any linear mapping (1.4.11) satisfying either one of the conditions (1.4.12) and (1.4.13) serves for our purpose as well as (1.4.10). The phase may still be defined by $\exp(i\delta_n) \equiv z_n$, but the relation (1.4.9) is no more valid.

In the case of (1.4.10), the value $u_n = \infty$ corresponds to $\delta_n = (2m + 1)\pi$, where m is an integer, while for (1.4.11) it corresponds to the phase angle of α/γ (mod 2π). In any case, the phase changes by just 2π as u_n runs from negative (positive) infinity to positive (negative) infinity. We call the transform (1.4.11) which satisfies the condition (1.4.12) or (1.4.13) *Cayley-type transform.*

1.5. Boundary Conditions

We have hitherto considered only the fixed boundary condition. The generalization to other conditions brings about no difficulty. It amounts to slightly modifying the matrix elements at the boundaries.

For $n = 1$ and $n = N$, i.e. for the boundaries in the direction of transfer, one has to replace $\mathbf{C}_1$ by $\mathbf{C}_1 + \mathbf{D}_0^T/\nu$ and $\mathbf{C}_N$ by $\mathbf{C}_N + \nu'\mathbf{D}_N$

in the difference equation (1.1.1) and the secular matrix (1.1.4), leaving the condition (1.1.2) unchanged. However, in the phase-theoretical treatment, in which the secular matrix does not appear explicitly, it is more convenient to take the following alternative procedure. Namely, we impose the conditions

$$\mathbf{v}_0 = \mathbf{v}_0, \quad \mathbf{v}_1 = \nu \mathbf{v}_0, \tag{1.5.1}$$

and

$$\mathbf{v}_{N+1} = \nu' \mathbf{v}_N, \tag{1.5.2}$$

instead of (1.1.2). Correspondingly, the initial condition for the matrix solution $\mathbf{V}_n$ must be replaced by

$$\mathbf{V}_1 = \nu \mathbf{I}, \quad \mathbf{V}_0 = \mathbf{I}, \quad \text{or} \quad \mathbf{U}_1 = \mathbf{I}/\nu. \tag{1.5.3}$$

Then the solution satisfying (1.5.1) is given by $\mathbf{V}_n \mathbf{v}_0$.

For $n = N$, we require that there should exist a vector $\mathbf{z}$ such that

$$\mathbf{V}_{N+1}\mathbf{z} = \nu' \mathbf{V}_N \mathbf{z}. \tag{1.5.4}$$

In terms of the state ratio matrix, the condition (1.5.4) is stated as follows: there should exist a vector $\mathbf{v}$ satisfying

$$\mathbf{U}_{N+1}^{-1}\mathbf{v} = \nu' \mathbf{v}, \tag{1.5.5}$$

since then (1.5.4) is valid for the vector $\mathbf{z} = \mathbf{V}_N^{-1}\mathbf{v}$. In other words, it is sufficient that $\mathbf{U}_{N+1}^{-1}$ has an eigenvalue ν' for the condition (1.5.2) to be satisfied. The vector $\mathbf{z}$ is the initial vector for which the final condition is fulfilled.

The values $\nu = 1$ and $\nu' = 1$ correspond to the free boundary conditions at $n = 1$ and $n = N$, respectively, while $\nu = \infty$ and $\nu' = 0$ correspond to the fixed boundary conditions at $n = 1$ and $n = N$ respectively.

For the cyclic boundary condition, we require that

$$\mathbf{V}_{N+1} = \mathbf{V}_1, \quad \mathbf{V}_N = \mathbf{V}_0, \tag{1.5.6}$$

or

$$\mathbf{X}_{N+1} = \mathbf{X}_1, \tag{1.5.7}$$

which means that there exists a vector $\mathbf{X}_1$ such that

$$(\mathbf{T}_{N;1} - \mathbf{I})\,\mathbf{X}_1 = \mathbf{0}. \tag{1.5.8}$$

The necessary and sufficient condition for the existence of such $\mathbf{X}_1$ is

$$|\mathbf{T}_{N;1} - \mathbf{I}| = 0. \tag{1.5.9}$$

When the matrix $\mathbf{T}_{N;1}$ is unimodular, this comes out

$$\text{trace } \mathbf{T}_{N;1} = 2\mathbf{I}. \tag{1.5.10}$$

The cyclic boundary condition cannot, however, be conveniently formulated in terms of the state ratio matrix $\mathbf{U}_n$.

For the y-direction, i.e. the direction transverse to that of transfer, we have to modify the matrix elements of $\mathbf{C}_n$ at $m = 1$ and $m = M$. As a typical example consider the case of (1.1.5). Let the boundary conditions be $v_{n1} = \nu v_{n0}$ and $v_{n,M+1} = \nu' v_{nM}$, as before. Then for $m = 1$, the first term of the right-hand side of (1.1.5) must be replaced by

$$K'_{n1;n0} v_{n1}(1 - \nu)/\nu, \tag{1.5.11}$$

while for $m = M$, the second term must be replaced by

$$K'_{nM;n,M+1} v_{nM}(\nu' - 1). \tag{1.5.12}$$

This amounts to replacing the (1,1) and (M, M) elements of the matrix $\mathbf{K}'_n$ by

$$-K'_{n1;n0}(1 - \nu)/\nu + K'_{n1;n2} + K_{n1;n-1,1} + K_{n1;n+1,1} \tag{1.5.13}$$

and

$$K'_{nM;n,M-1} - K'_{nM;n,M+1}(\nu' - 1) + K_{nM;n-1,M} + K_{nM;n+1,M} \tag{1.5.14}$$

respectively. As before, $\nu' = \nu = 1$ corresponds to the free boundary condition and $\nu = \infty$ and $\nu' = 0$ to the fixed boundary conditions for $m = 1$ and for $m = M$ respectively.

For the cyclic boundary condition we have to add to the matrix $\mathbf{K}'_n$ the $(1, M)$ element $K_{n1;n0}$ and the $(M, 1)$ element $K_{nM;n,M+1}$.

Every regular (periodic) system can be described by only one kind of transfer matrix, owing to the fact mentioned in § 1.3 that an arbitrary product of transfer matrices can be regarded as a single transfer matrix. Even when the system is not periodic, it may contain the *regular parts* which can be described by a single transfer matrix (*regular transfer matrix*). In such cases it is convenient to use the representation in which the *regular submatrix* $\mathbf{L}_n \equiv \mathbf{D}_n^{-1}\mathbf{C}_n$ or $\mathbf{K}_n^{-1}(\mathbf{K}'_n - \lambda\mathbf{M}_n)$ appearing in the regular transfer matrix is diagonalized.

Consider, for example, a regular square lattice in which both the mass and force constants are constant so that one may drop all the suffices n and m from the vectors and matrices. The matrices

M, **K**, and **K**′ then become $m\mathbf{I}$, $K\mathbf{I}$, and

$$\mathbf{K}' = \begin{pmatrix} 2(K+K') - K'/\nu & -K' & & & (-K') \\ -K' & 2(K+K') & -K' & & \\ & -K' & \ddots & & \\ & & & & -K' \\ (-K') & & & -K' & 2(K+K') - \nu' K' \end{pmatrix} \tag{1.5.15}$$

respectively. Therefore the regular submatrix $\mathbf{L} = \mathbf{K}^{-1}(\mathbf{K}' - \omega^2\mathbf{M})$ is

$$\mathbf{L} = \begin{pmatrix} 2+\gamma(2-\nu^{-1}) - m\omega^2/K & -\gamma & & (-\gamma) \\ -\gamma & 2(1+\gamma) - m\omega^2/K & & \\ & & \ddots & \\ & & & -\gamma \\ (-\gamma) & & -\gamma & 2+\gamma(2-\nu') - m\omega^2/K \end{pmatrix}, \tag{1.5.16}$$

where $\gamma \equiv K'/K$. In (1.5.15) and (1.5.16), the elements in the brackets should be omitted except in the case of cyclic boundary condition, for which one should put $\nu = \infty$ and $\nu' = 0$.

Since the eigenvalues and eigenvectors of a matrix of the type (1.5.16) are well known, we omit here the derivation and only enumerate the results. For the fixed boundary condition, i.e. for $\nu = \infty$ and $\nu' = 0$, the eigenvalues and normalized eigenfunctions of **L** are

$$\chi_\mu = \Omega - 2\gamma \cos 2\theta_\mu, \quad \theta_\mu = \mu\pi/2(M+1), \quad \mu = 1, 2, \ldots, M \tag{1.5.17}$$

and

$$Z_m^{(\mu)} = \left(\frac{1}{M+1}\right)^{1/2} \sin 2m\theta_\mu \tag{1.5.18}$$

respectively, where

$$\Omega \equiv 2(1+\gamma) - m\omega^2/K. \tag{1.5.19}$$

For the free boundary condition, i.e. for $\nu = \nu' = 1$, they become

$$\chi_\mu = \Omega - 2\gamma \cos 2\theta_\mu, \quad \theta_\mu = \mu\pi/2M, \quad \mu = 0, 1, \ldots, M-1, \tag{1.5.20}$$

and

$$Z_m^{(\mu)} = \left(\frac{2 - \delta_{\mu,0}}{M}\right)^{1/2} \cos(2m - 1)\,\theta_\mu \tag{1.5.21}$$

respectively. Finally, for the cyclic boundary condition, we have

$$\chi_\mu = \Omega - 2\gamma \cos 2\theta_\mu, \quad \theta_\mu = \mu\pi/M, \quad \mu = 1, 2, \ldots, M, \tag{1.5.22}$$

and

$$Z_m^{(\mu)} = \left(\frac{1}{M}\right)^{1/2} \exp(2im\theta_\mu). \tag{1.5.23}$$

If we put

$$m\omega^2 = 4K \sin^2 \beta_\mu + 4K' \sin^2 \theta_\mu, \tag{1.5.24}$$

the eigenvalues can be written simply as

$$\chi_\mu = 2 \cos 2\beta_\mu \tag{1.5.25}$$

in each of above three cases. The quantities β_μ, as well as ω^2, play the role of energy parameters.

Corresponding formulae can easily be obtained also for three-dimensional lattices.

1.6. Regular Lattices

For later reference, we calculate in this section the eigenfrequencies of some regular vibrational systems. First let us consider a regular square lattice. The regular transfer matrix is

$$\mathbf{T} = \begin{pmatrix} \mathbf{0} & \mathbf{I} \\ -\mathbf{I} & \mathbf{L} \end{pmatrix}, \tag{1.6.1}$$

the eigenvalues of which are

$$\boldsymbol{\theta}_\pm = \tfrac{1}{2}\{\mathbf{L} \pm (\mathbf{L}^2 - 4\mathbf{I})^{1/2}\}. \tag{1.6.2}$$

In the representation which diagonalizes $\mathbf{T}$, the equation

$$\mathbf{X}_{N+1} = \mathbf{T}^N \mathbf{X}_1 \tag{1.6.3}$$

becomes

$$(\boldsymbol{\theta}_- - \boldsymbol{\theta}_+)^{-1} \begin{pmatrix} \boldsymbol{\theta}_- \mathbf{V}_N - \mathbf{V}_{N+1} \\ -\boldsymbol{\theta}_+ \mathbf{V}_N + \mathbf{V}_{N+1} \end{pmatrix} = (\boldsymbol{\theta}_- - \boldsymbol{\theta}_+)^{-1} \begin{pmatrix} \boldsymbol{\theta}_+^N & \mathbf{0} \\ \mathbf{0} & \boldsymbol{\theta}_-^N \end{pmatrix} \begin{pmatrix} \boldsymbol{\theta}_- \mathbf{V}_0 - \mathbf{V}_1 \\ -\boldsymbol{\theta}_+ \mathbf{V}_0 + \mathbf{V}_1 \end{pmatrix}. \tag{1.6.4}$$

Let us consider here the case of free boundary condition only. From (1.5.3) and (1.5.4) it turns out that there must exist a vector **z** such that

$$(\boldsymbol{\theta}_+^N - \boldsymbol{\theta}_-^N)\,\mathbf{z} = \mathbf{0}. \tag{1.6.5}$$

Thus we must have

$$\det(\boldsymbol{\theta}_+^N - \boldsymbol{\theta}_-^N) = 0, \tag{1.6.6}$$

which comes out to be

$$\sin 2N\beta_\mu = 0, \tag{1.6.7}$$

since from (1.5.25) and (1.6.2) the eigenvalues of $\boldsymbol{\theta}_\pm$ are given by $\exp(\pm 2i\beta_\mu)$. Then by (1.5.20) and (1.5.24) we obtain the eigenfrequencies

$$\omega_{\varrho\mu}^2 = (4K/m)\sin^2(\varrho\pi/2N) + 4(K'/m)\sin^2(\mu\pi/2M),$$

$$\varrho = 0, 1, 2, \ldots, N-1; \quad \mu = 0, 1, 2, \ldots, M-1. \tag{1.6.8}$$

The maximum eigenfrequency is therefore given by

$$\omega_{\max}^2 = 4(K+K')/m. \tag{1.6.9}$$

The spectral density function $D(\omega^2)$ corresponding to (1.6.8) has been calculated analytically by Montroll (1955). It is illustrated in Fig. 1.4.

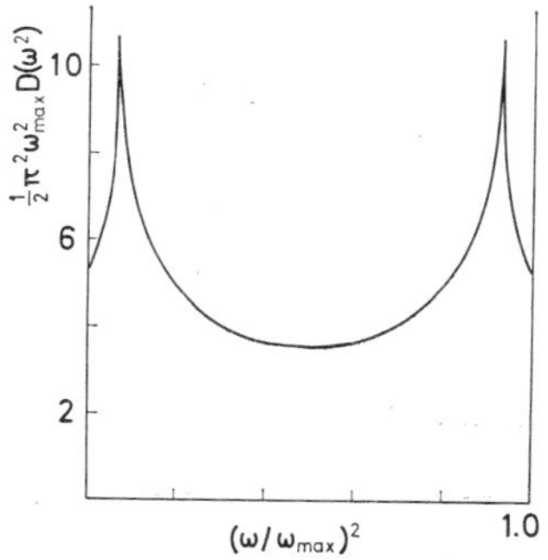

FIG. 1.4. Density of the squared-frequency spectrum of a two-dimensional lattice with $K'/K = 1/9$. (After Montroll (1955).)

In the one-dimensional case (i.e. for the regular monatomic chain), (1.6.8) reduces to

$$\omega_\mu^2 = (4K/m)\sin^2(\mu\pi/2N), \quad \mu = 0, 1, 2, \ldots, N-1, \tag{1.6.10}$$

which gives the maximum frequency

$$\omega_{max}^2 = 4K/m. \tag{1.6.11}$$

The corresponding spectral density is easily found to be

$$D(\omega^2) = \frac{2}{\pi\omega(\omega_{max}^2 - \omega^2)^{1/2}}, \tag{1.6.12}$$

which has a U-shaped form with the singularities at $\omega^2 = 0$ and ω_{max}^2, as shown in Fig. 1.5.

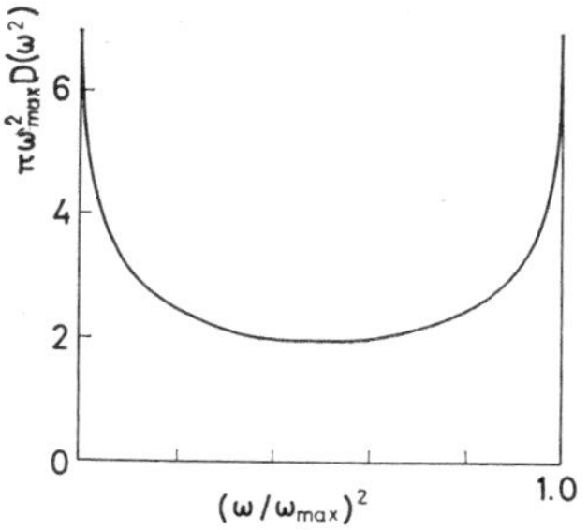

FIG. 1.5. Density of the squared-frequency spectrum of a one-dimensional chain. (After Montroll (1955).)

From (1.6.4) we see that in the representation which diagonalizes **T**, the phase of the components of state vector increases or decreases by just 2β during a transfer from an atom to the next. Thus the quantity β, which plays the role of energy parameter in the one-dimensional case, has the meaning of wave number. We shall occasionally call it *wave-number parameter*.

Next consider a linear diatomic chain composed of atoms with masses $m^{(0)}$ and $m^{(1)} (m^{(0)} < m^{(1)})$. Assume that the atoms with mass $m^{(0)}$ are situated at $n = 1, 3, 5, \ldots$, and the others are at $n = 2, 4, \ldots$ and denote $K_{2l-1,2l}$ and $K_{2l,2l+1}$ (l: integer) by K_0 and K_1 respectively. Then, according to (1.3.7), the transfer matrices which transfer $\mathbf{X}_{2l-1}$ to $\mathbf{X}_{2l}$ and $\mathbf{X}_{2l}$ to $\mathbf{X}_{2l+1}$ are

$$\mathbf{T}^{(0)} \equiv \begin{pmatrix} 0 & 1 \\ -K_1/K_0 & (K_1 + K_0 - \omega^2 m^{(0)})/K_0 \end{pmatrix} \tag{1.6.13}$$

and

$$\mathbf{T}^{(1)} \equiv \begin{pmatrix} 0 & 1 \\ -K_0/K_1 & (K_0 + K_1 - \omega^2 m^{(1)})/K_1 \end{pmatrix} \tag{1.6.14}$$

respectively, and their product is

$$\mathbf{T} \equiv \mathbf{T}^{(1)}\mathbf{T}^{(0)}$$

$$= \begin{pmatrix} -\dfrac{K_1}{K_0} & \dfrac{K_1 + K_0 - \omega^2 m^{(0)}}{K_0} \\ -\dfrac{K_0 + K_1 - \omega^2 m^{(1)}}{K_0} & -\dfrac{K_0}{K_1} + \dfrac{(K_1 + K_0 - \omega^2 m^{(0)})(K_1 + K_0 - \omega^2 m^{(1)})}{K_0 K_1} \end{pmatrix}. \quad (1.6.15)$$

The eigenvalues of **T** are easily found to be given by $\theta_{\pm} = \exp(\pm 2i\beta)$, if we put

$$4 \sin^2\beta = \{\omega^2(K_0 + K_1)(m^{(0)} + m^{(1)}) - \omega^4 m^{(0)} m^{(1)}\}/K_0 K_1 . \quad (1.6.16)$$

Thus the eigenfrequencies are obtained by solving (1.6.16) for $\beta = \mu\pi/2N$. The result is

$$\omega_\mu^2 = \frac{1}{2m^{(0)}m^{(1)}} \left[(m^{(0)} + m^{(1)})(K_0 + K_1) \pm \left\{ (m^{(0)} + m^{(1)})^2 (K_0 + K_1)^2 - 16 K_0 K_1 m^{(0)} m^{(1)} \sin^2\left(\frac{\mu\pi}{2N}\right) \right\}^{1/2} \right],$$

$$\mu = 0, 1, 2, \ldots, N - 1. \quad (1.6.17)$$

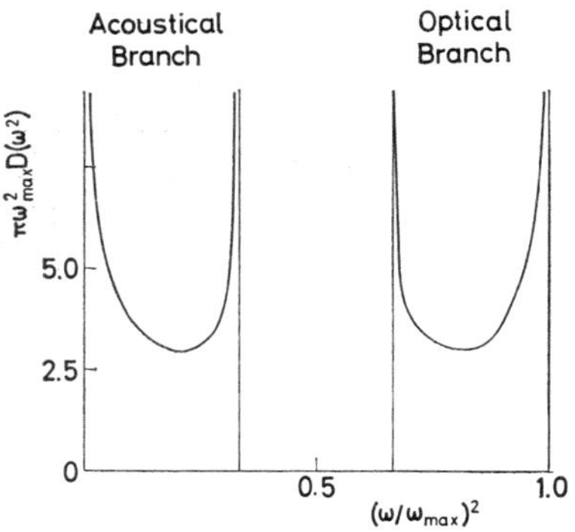

FIG. 1.6. Density of the squared-frequency spectrum of a diatomic chain with mass ratio 2.

This is the well-known result, showing that there appear two branches of eigenfrequencies (*acoustical and optical branches*). The corresponding spectral density $D(\omega^2)$ is shown in Fig. 1.6.

It is not necessary to reproduce the results for other boundary conditions, since the calculation runs in an entirely similar manner, leading to very similar formulae, and, moreover, the spectral density $D(\omega^2)$ does not depend on the boundary condition.

CHAPTER 2

Numerical Methods and Calculated Spectra

2.1. Method of Negative Factor Counting

The secular determinant $|\mathbf{S}(\lambda)|$ can naturally be factorized as

$$|\mathbf{S}(\lambda)| = \prod_{\mu=1}^{\mathbf{N}} (\lambda_\mu - \lambda), \tag{2.1.1}$$

where $\mathbf{N}$ is the total number of eigenvalues. Let us introduce the characteristic function defined by

$$\begin{aligned} \Gamma(\lambda - t) &= \frac{1}{\mathbf{N}} \lim_{\mathbf{N}\to\infty} \log |\mathbf{S}(\lambda - t)| \\ &= \frac{1}{\mathbf{N}} \lim_{\mathbf{N}\to\infty} \sum_{\mu=1}^{\mathbf{N}} \log (\lambda_\mu - \lambda + t) \\ &= \int \log (\mu - \lambda + t)\, D(\mu)\, d\mu. \end{aligned} \tag{2.1.2}$$

From the well-known property of the logarithmic function one obtains the relation

$$\mathrm{Re}\left\{\frac{1}{i\pi} \Gamma(\lambda - i0)\right\} = \int_{-\infty}^{\lambda} D(\mu)\, d\mu \equiv M(\lambda), \tag{2.1.3}$$

where $M(\lambda)$ is the fraction of eigenvalues less than λ (the *integrated density* or *integrated spectrum*).

Imagine now that $|\mathbf{S}(\lambda)|$ can be factorized in some other way:

$$|\mathbf{S}(\lambda)| = \prod_{\mu=1}^{\mathbf{N}} Q_\mu(\lambda), \tag{2.1.4}$$

where each factor $Q_\mu(\lambda)$ is a function of λ. Then $\Gamma(\lambda - t)$ can be written

$$\Gamma(\lambda - t) = \frac{1}{\mathbf{N}} \lim \sum_{\mu=1}^{\mathbf{N}} \log Q_\mu(\lambda - t) = \int_{-\infty}^{+\infty} P_{\lambda-t}(Q) \log Q dQ. \tag{2.1.5}$$

Here $P_{\lambda-t}(Q)$ is the density distribution function of the quantity Q_μ. Defining $q = Q(\lambda) + \lambda$ and substituting the expression (2.1.5) into (2.1.3), we have

$$M(\lambda) = \mathrm{Re}\left\{\frac{1}{i\pi} \int_{-\infty}^{\infty} P_{\lambda-i0}(q) \log (q - \lambda + i0)\, dq\right\}. \tag{2.1.6}$$

If we assume that $P_\lambda(q)$ is analytic on the real axis as a function of the energy parameter, we can write $P_\lambda(q)$ instead of $P_{\lambda-i0}(q)$, so that we get

$$M(\lambda) = \mathrm{Re}\left\{\frac{1}{i\pi} \int_{-\infty}^{\infty} P_\lambda(q) \log (q - \lambda + i0)\, dq\right\}$$

$$= \int_{-\infty}^{\infty} P_\lambda(Q) \log (Q + i0)\, dQ = \int_{-\infty}^{\infty} P_\lambda(Q)\, dQ. \tag{2.1.7}$$

This result can be interpreted that the number of the quantities q less than λ, or the number of negative factors Q_μ, must be equal to the number of eigenvalues less than λ. Thus the spectrum of the system can be calculated by counting the number of negative factors Q_μ.

To find a convenient factorization of $|\mathbf{S}(\lambda)|$ is the next task. For this purpose we use the fact that any tridiagonal matrix may be decomposed into a product of two triangular matrices of simple form. Precisely, one can put

$$\mathbf{S}(\lambda) = \mathbf{L}(\lambda)\, \mathbf{U}(\lambda), \tag{2.1.8}$$

where $\mathbf{L}$ and $\mathbf{U}$ are the matrices of order $\mathbf{N}(= N \times M)$:

$$\mathbf{L}(\lambda) = \begin{pmatrix} \mathbf{I} & \mathbf{0} & & & \\ \mathbf{L}_2 & \mathbf{I} & \mathbf{0} & & \\ & \mathbf{L}_3 & \mathbf{I} & & \\ & & & \ddots & \\ & & & \mathbf{L}_N & \mathbf{I} \end{pmatrix}, \tag{2.1.9}$$

$$\mathbf{U}(\lambda) = \begin{pmatrix} \mathbf{U}_1 & \mathbf{V}_1 & & & \\ \mathbf{0} & \mathbf{U}_2 & \mathbf{V}_2 & & \\ & \mathbf{0} & \mathbf{U}_3 & \mathbf{V}_3 & \\ & & & \ddots & \\ & & & \mathbf{0} & \mathbf{U}_N \end{pmatrix}. \tag{2.1.10}$$

In both matrices the partitioning corresponds to that of $\mathbf{S}(\lambda)$.

From (2.1.8) one obtains the relations

$$\begin{aligned} \mathbf{C}_1 &= \mathbf{U}_1, \\ \mathbf{C}_n &= \mathbf{L}_n\mathbf{V}_n + \mathbf{U}_n, \\ \mathbf{D}_{n-1} &= \mathbf{V}_n, \\ \mathbf{D}_{n-1}^T &= \mathbf{L}_n\mathbf{U}_{n-1}, \quad n = 2, 3, \ldots, \end{aligned} \tag{2.1.11}$$

which lead to the recurrence equation

$$\mathbf{U}_n = \mathbf{C}_n - \mathbf{D}_{n-1}^T\mathbf{U}_{n-1}^{-1}\mathbf{D}_{n-1}, \quad \mathbf{U}_1 = \mathbf{C}_1. \tag{2.1.12}$$

This equation is similar to (1.4.3). In fact, defining $\mathbf{U}_n$ by

$$\mathbf{U}_n \equiv -\mathbf{D}_n\mathbf{V}_{n+1}\mathbf{V}_n^{-1}, \tag{2.1.13}$$

we readily get (2.1.12) from (1.3.2). Thus $\mathbf{U}_n$ is essentially the state ratio matrix introduced in § 1.4, and the recurrence equation (2.1.12) can be regarded as a special case of the general transformation (1.4.1).

Let us denote the eigenvalues of the matrix $\mathbf{U}_n$ by u_{nl} ($l = 1, 2, \ldots, M$). Then it is clear that

$$|\mathbf{S}(\lambda)| = \prod_{nl} u_{nl}. \tag{2.1.14}$$

This is the required factorization, so that the number of negative eigenvalues u_{nl} gives the number of eigenvalues λ_μ less than λ. This statement was derived by Dean and Martin (1960b) and called *negative-eigenvalue theorem*. The method of *negative factor counting* is based on this theorem, and amounts to counting the number of negative u_{nl}.

In the one-dimensional case, the matrices $\mathbf{U}_n$, $\mathbf{C}_n$, and $\mathbf{D}_n$ are reduced to the scalars u_n, C_n, and D_n respectively, so that (2.1.12) and (2.1.14) become

$$u_n = C_n - D_{n-1}^2/u_{n-1} \tag{2.1.15}$$

and

$$|\mathbf{S}(\lambda)| = u_1 u_2 \dots u_N \tag{2.1.16}$$

respectively. In this case the number of negative u_n gives the number of eigenvalues less than λ. As was mentioned above, u_n is almost the same thing as the state ratio. It will be shown in Chapter 3 that the negative-eigenvalue theorem, or its slight generalization, is valid for the state ratio in every case.

The method of negative factor counting is particularly suitable for the numerical computation of the spectrum. The so-called *node-counting method*, which has been often used for numerical computation of the spectrum of the one-dimensional electronic system, is intimately connected with the method of negative factor counting. The relation between the two methods can be seen from (1.3.18), for it shows that in the electronic case the quantity ψ_n/ψ_n' may play the role of the state ratio. A bit of consideration shows that the number of intervals in which ψ/ψ' is negative is equal to the number of nodes of the solution of the wave equation, which in turn is known to give the number of eigenvalues less than λ. It should also be remarked that, in the one-dimensional case, the negative-eigenvalue theorem may also be considered as a consequence of the well-known theorem of Sturm, the usefulness of which for the calculation of frequency spectrum was pointed out by Rosenstock and McGill (1962).

2.2. Method of Functional Equation

Although the method of negative factor counting is very useful for computing the spectrum numerically, it has a defect that one can calculate the spectrum only for individual samples of the disordered system. Such samples are, moreover, necessarily finite. This may give rise to an extra error besides the purely numerical one. Alternatively, we may calculate directly the averaged spectrum for the infinite system by setting up a functional equation for the distribution function $P_\lambda(Q)$ and solving it numerically. Such a method will give the results which are subject only to the numerical error in solving the equation.

Consider, for example, the recurrence equation (2.1.15) for u_n, and assume that the pair (C_n, D_{n-1}) takes a value $(C^{(i)}, D^{(j)})$ with the probability $P^{(i,j)}(i, j = 1, 2, \ldots)$. Equation (2.1.15) can be rewritten as

$$u_{n-1}(\lambda) = D_{n-1}^2/(C_n - u_n(\lambda)), \tag{2.2.1}$$

from which it is readily seen that $P_\lambda(u)$ must satisfy the equation

$$P_\lambda(u)\,du = \sum_{(i,j)} P^{(i,j)} P_\lambda\left[\frac{D^{(j)2}}{(C^{(i)} - u)}\right] d\left[\frac{D^{(j)2}}{(C^{(i)} - u)}\right], \tag{2.2.2}$$

or

$$P_\lambda(u) = \sum_{(i,j)} P^{(i,j)} \frac{D^{(j)2}}{(C^{(i)} - u)^2} P_\lambda\left[\frac{D^{(j)2}}{(C^{(i)} - u)}\right]. \tag{2.2.3}$$

This is a functional equation for the density $P_\lambda(u)$. In the above derivation it has been assumed implicitly that in an infinite system $P_\lambda(u)$ is stationary, i.e. it does not depend on n. Taking this into account, we can obtain the fraction of negative u_n, and consequently the fraction of the eigenvalues less than λ, by integrating $P_\lambda(u)$ from negative infinity to zero. Also the characteristic function $\Gamma(\lambda)$ can be obtained from $P_\lambda(u)$ through (2.1.5).

It is almost evident that similar equations can be derived for the state ratio in every case, inasmuch as the successive state ratios u_n and u_{n+1} are related to each other through the linear transformation (1.4.2). It is also possible to construct in a similar way the functional equation for the phase δ_n.

Such equations were considered by many authors. In his pioneering work on the frequency spectrum of disordered chain, Dyson (1953) derived for the first time an integral equation corresponding to (2.2.3) on the basis of the concept of characteristic function. Since then, several different versions and generalizations have been presented. Bellman (1956) made Dyson's involved theory much simpler and understandable. Dean (1956, 1959), Dean and Martin (1960a), and Englman (1958) generalized the argument to the case of general secular matrix, and obtained the functional equations which replace Dyson's integral equation. Schmidt (1957) presented a somewhat different treatment of the problem, attaining a functional equation which is a special case of (2.2.3). Des Cloizeaux (1957) rederived Dyson's result by a method different from those of Dyson and Bellman. Borland (1961a) derived an integral equation for the case of an electron moving in a one-dimensional array of potentials. The latter three theories are based on the property of the phase and intimately connected with the phase theory.

Schlup (1963b) gave a power-series expansion in the concentration of impurity of the solution of Schmidt's equation for the case of isotopically disordered chain. Hori (1962) presented the formal method (used above) of deriving the negative-eigenvalue theorem, and discussed the connection between these various theories.

Unfortunately, it is difficult in general to obtain the solution of these equations in a closed form, so that one is obliged to have recourse to either approximate or numerical solution. Up to the present, the approximate solutions have scarcely given us any significant knowledge on the detailed structure of the spectrum, while the numerical solutions have provided us with a good deal of information concerning the spectrum.

2.3. Method of Green Function

Another method of calculating the spectrum can be derived by starting again from the characteristic function $\Gamma(\lambda - t)$ defined by (2.1.2). Differentiation of (2.1.2) with respect to t gives

$$\Gamma_t(\lambda - t) = \int D(\mu)\, d\mu/(\mu - \lambda + t)$$
$$= \frac{1}{\mathbf{N}} \lim \sum_{\mu=1}^{\mathbf{N}} 1/(\lambda_\mu - \lambda + t). \tag{2.3.1}$$

Using the formula

$$\delta(x) = \frac{1}{\pi} \lim_{\varepsilon\to 0+} \mathrm{Im}\,\{1/(x - i\varepsilon)\}, \tag{2.3.2}$$

one obtains

$$\lim_{\varepsilon\to 0+} \frac{1}{\pi} \mathrm{Im}\, \Gamma_t(\lambda + i\varepsilon) = \lim \frac{1}{\pi} \mathrm{Im} \int D(\mu)\, d\mu/(\mu - \lambda - i\varepsilon)$$
$$= \int \delta(\mu - \lambda)\, D(\mu)\, d\mu = D(\lambda). \tag{2.3.3}$$

Thus the spectral density $D(\lambda)$ is obtained if we can calculate in some way the function $\Gamma_t(\lambda - t)$ without knowing in advance the values of λ_μ. This can be done as follows.

For the sake of simplicity we consider only the one-dimensional case. The generalization to the higher-dimensional case is straightforward. Consider the one-dimensional equation corresponding to (1.1.20):

$$(A_n - \lambda)\, w_n - B_n w_{n+1} - B_{n-1} w_{n-1} = 0 \tag{2.3.4}$$

and the associated inhomogeneous equation

$$(A_n - \lambda)\, G_{nl}(\lambda) - B_n G_{n+1,\,l}(\lambda) - B_{n-1} G_{n-1,\,l}(\lambda) = \delta_{nl}. \tag{2.3.5}$$

These can be put in the matrix form

$$(\mathbf{P} - \lambda \mathbf{I})\, \mathbf{w} = \mathbf{0} \tag{2.3.6}$$

and

$$(\mathbf{P} - \lambda \mathbf{I})\, \mathbf{G}(\lambda) = \mathbf{I}, \tag{2.3.7}$$

where $\mathbf{w} \equiv (w_1, w_2, \ldots, w_N)^T$ and $\mathbf{G}(\lambda)$ is the matrix with the elements $G_{nl}(\lambda)$. The formal solution of (2.3.7) (the *Green operator*) is

$$\mathbf{G}(\lambda) = (\mathbf{P} - \lambda \mathbf{I})^{-1} = \sum_\mu \mathbf{w}^{(\mu)} \mathbf{w}^{(\mu)T} / (\lambda_\mu - \lambda), \tag{2.3.8}$$

where $\mathbf{w}^{(\mu)}$ is the normalized eigenvector of $\mathbf{P}$, or the solution of the homogeneous equation corresponding to the eigenvalue λ_μ. The elements $G_{nl}(\lambda)$ (the *Green function*) are given by

$$G_{nl}(\lambda) = \sum_\mu w_n^{(\mu)} w_l^{(\mu)} / (\lambda_\mu - \lambda), \tag{2.3.9}$$

where $\mathbf{w}_n^{(\mu)}$ is the nth component of $\mathbf{w}^{(\mu)}$.

Summing up (2.3.9) with respect to n, and using the normalization condition of the eigenvector $\mathbf{w}^{(\mu)}$, we obtain

$$\text{trace}\ \mathbf{G}(\lambda) \equiv \sum_n G_{nn}(\lambda) = \sum_\mu 1/(\lambda_\mu - \lambda). \tag{2.3.10}$$

Thus we reach

$$\Gamma_t(\lambda + i\varepsilon) = \frac{1}{N} \lim \sum_n G_{nn}(\lambda + i\varepsilon), \tag{2.3.11}$$

and

$$\lim_{\varepsilon \to 0+} \frac{1}{\pi} \operatorname{Im} \lim \frac{1}{N} \sum_n G_{nn}(\lambda + i\varepsilon) = D(\lambda). \tag{2.3.12}$$

By the formula (2.3.2) we can also get the relation

$$2\pi \sum_\mu \delta(\lambda_\mu - \lambda)\, w_n^{(\mu)} w_n^{(\mu)} = 2 \lim_{\varepsilon \to 0+} \operatorname{Im} G_{nn}(\lambda + i\varepsilon). \tag{2.3.13}$$

Integrating this over a short interval $(\lambda_\mu - \Delta) < \lambda < (\lambda_\mu + \Delta)$, containing only the eigenvalue λ_μ and none other, one obtains

$$\lim_{\varepsilon \to 0+} \int_{\lambda_\mu - \Delta}^{\lambda_\mu + \Delta} \operatorname{Im} G_{nn}(\lambda + i\varepsilon)\, d\lambda = \pi |w_n^{(\mu)}|^2, \tag{2.3.14}$$

which gives the distribution of intensity in the μth normal mode.

Thus we see that to get the function $\Gamma_t(\lambda - t)$ amounts to obtaining the Green function, and, moreover, that one can obtain therefrom not only the spectral density but also the intensity distribution in each normal mode. Calculation of the Green function can be carried out, for example, as follows. Let $y_n^{(1)}(\lambda)$ and $y_n^{(2)}(\lambda)$ be the solutions of the homogeneous equation (2.3.4), each of which satisfies the boundary condition at one of the ends:

$$y_1^{(1)}(\lambda) = y_N^{(2)}(\lambda) = 1, \tag{2.3.15}$$

and construct the function

$$G_{nl}(\lambda) \equiv \begin{cases} Cy_n^{(1)}(\lambda)\, y_l^{(2)}(\lambda), & n \leqq l, \\ Cy_n^{(1)}(\lambda)\, y_l^{(2)}(\lambda), & n \geqq l. \end{cases} \tag{2.3.16}$$

For $n \neq l$, it is clear that $G_{nl}(\lambda)$ is a solution of (2.3.5). For $n = l$, substitute (2.3.16) into (2.3.5) and put $n = 1$. Then using the relation

$$(A_1 - \lambda)\, y_1^{(1)}(\lambda) - B_1 y_2^{(1)}(\lambda) = 0, \tag{2.3.17}$$

we get

$$C = 1/(B_1[y_2^{(1)}(\lambda)\, y_1^{(2)}(\lambda) - y_2^{(2)}(\lambda)\, y_1^{(1)}(\lambda)]), \tag{2.3.18}$$

which is a constant independent of n, since the expression in the square bracket is just the Wronskian of $y_n^{(1)}$ and $y_n^{(2)}$. Further it may readily be verified that the symmetry relation

$$G_{kl} = G_{lk} \tag{2.3.19}$$

holds. We can use the formula (2.3.16) for the numerical computation of $G_{nl}(\lambda)$, and hence the quantities such as $D(\lambda)$ and $|w_n^{(\mu)}|^2$.

The method of Green function is not only useful for the numerical calculation, but also serves as an effective tool for investigating the impurity modes associated with isolated or extended impurities, and as a starting point of various approximate procedures of investigating the spectrum.

2.4. Frequency Spectra of Disordered Lattices

By using the method of negative factor counting, Dean (1960, 1961) calculated numerically the vibrational frequency spectra of various linear disordered chains which contain in equal concentration two isotopes with masses $m^{(0)}$ and $m^{(1)}$ ($m^{(0)} < m^{(1)}$). Figures 2.1–2.4 show the spectra of such lattices with $N = 8000$

for the case of the complete disorder and for mass ratios $m^{(1)}/m^{(0)} = 5/4, 3/2, 2$ and 3. It is seen that, for the large mass ratio, the spectrum has a remarkably fine structure with many distinct peaks in high-frequency region. The peaks already exist for small mass ratio, but become particularly numerous and conspicuous for $m^{(1)}/m^{(0)} \geqq 2$. It is important to remark that for these mass ratios there appear deep valleys between peaks in which the spectral density seems to vanish, and that these valleys correspond to particular rational values of β/π, as indicated in Fig. 2.3. As the mass ratio increases, the valleys become wider and the peaks become acuter.

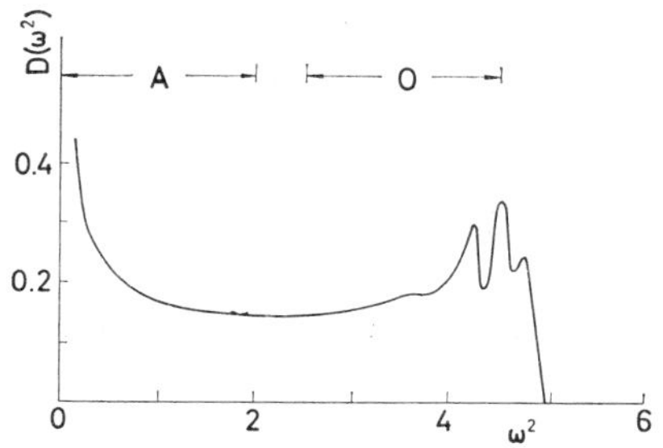

FIG. 2.1. The squared-frequency spectrum of the isotopically disordered chain with mass ratio 5/4 and length 8000. The intervals indicated by *A* and *O* are the regions of acoustical and optical branches, respectively, of the corresponding regular lattice. (After Dean (1960); Crown copyright reserved.)

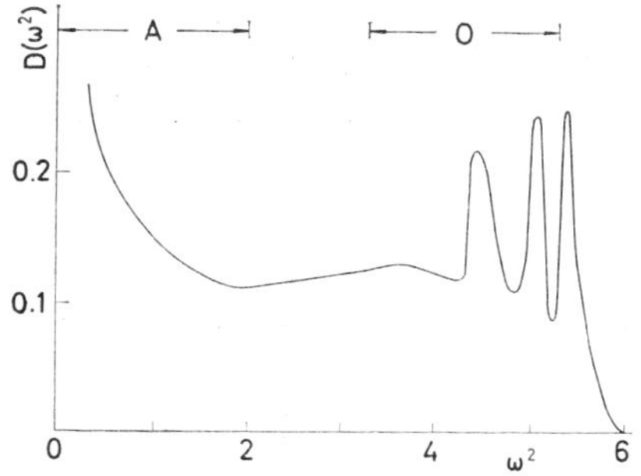

FIG. 2.2. The squared-frequency spectrum of the isotopically disordered chain with mass ratio 3/2. (After Dean (1960); Crown copyright reserved.)

This result obtained by Dean gave rise to a considerable surprise, since it had been believed generally that the effect of disorder on the spectrum should be to produce a fairly diffuse distribution of frequencies with little or no detailed structure. Even after Dean's

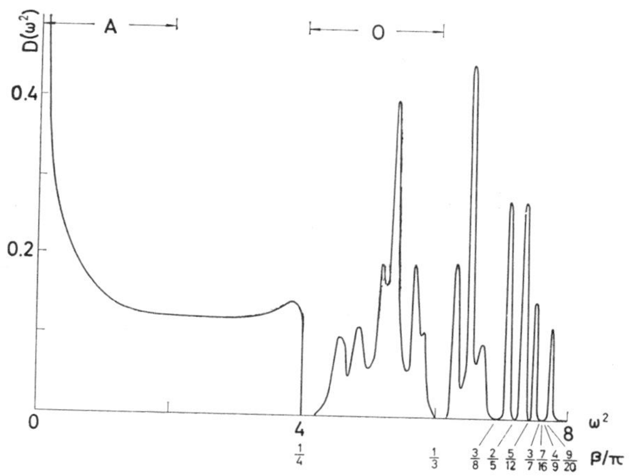

FIG. 2.3. The squared-frequency spectrum of the isotopically disordered chain with mass ratio 2. The deep valleys correspond to particular rational values of β/π. (After Dean (1960); Crown copyright reserved.)

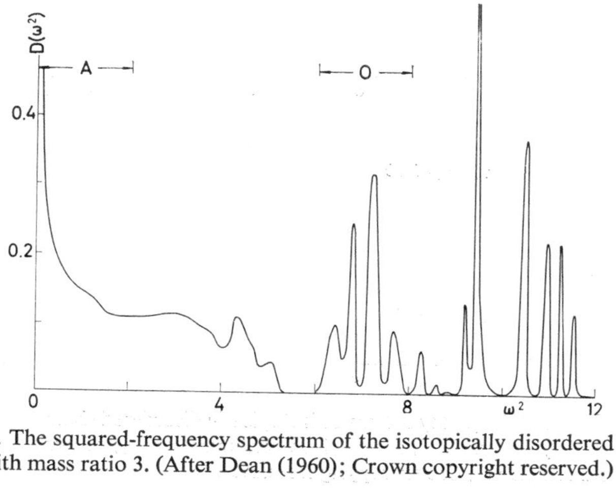

FIG. 2.4. The squared-frequency spectrum of the isotopically disordered chain with mass ratio 3. (After Dean (1960); Crown copyright reserved.)

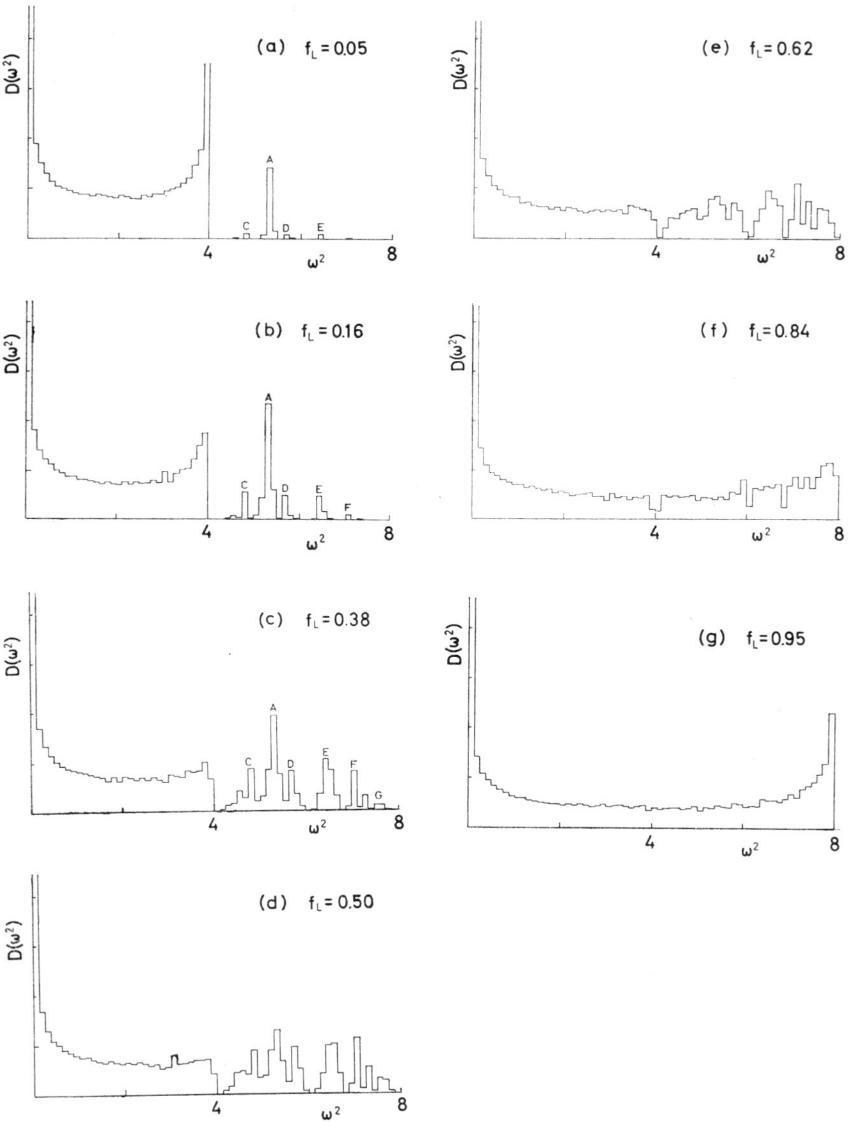

FIG. 2.5. The squared-frequency spectra of isotopically disordered chains with mass ratio 2, length 8000, and various concentrations f_L of light atoms. (After Dean (1961); Crown copyright reserved.)

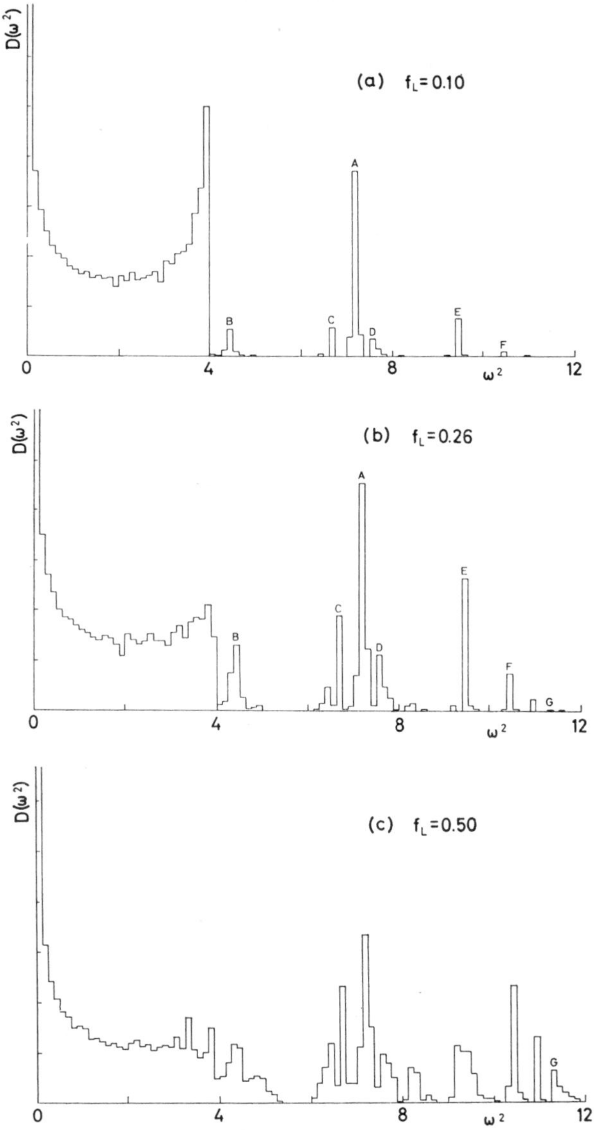

FIG. 2.6. The squared-frequency spectra of isotopically disordered chains with mass ratio 3 and concentrations f_L = 0·10, 0·26, and 0.50. (After Dean (1961); Crown copyright reserved.)

paper appeared, many investigators continued to cast a doubt upon his result on the ground that Dean's calculation was numerical and could not be completely free from the errors due to the finiteness of the chains used in the calculation or other computational errors. Such a doubt has now been cleared up, however, by the empirical explanation of the peaks given by Dean himself, which will be mentioned below, and the theoretical explanation of both the valleys and peaks given by Matsuda (1964), Hori (1964), and Hori and Fukushima (1963), which will be described in detail in the following chapters.

Dean investigated also how the spectrum is affected by the change of concentration ratio of two isotopes. Figures 2.5 and 2.6 show his results for mass ratios 2 and 3 respectively. The length of the chain was $N = 8000$ for both cases. When the concentration of light atom f_L is small, the spectrum is similar to that of the monatomic chain of heavy atoms given by (1.6.12). The differences are that the singularity at the top of the frequency band is less intense and that a number of peaks exist outside the band in the present case. As f_L increases, the maximum at the top of the main band becomes lower and the high-frequency peaks grow at the expense of the main band. At about $f_L \sim 1/6$, the peak system appears most clearly: There is one dominant peak, with several peaks of minor intensity. As f_L increases further, the primary peak begins to decrease in size, while the other peaks continue to grow. At $f_L = 0{\cdot}5$, all peaks become roughly of comparable intensity. Beyond $f_L = 0{\cdot}5$, the peaks broaden and merge into one another, although the peak structure persists in high-frequency region up to quite high values of $f_L(\sim 0{\cdot}8)$. Finally, as $f_L \to 1$, only an intense peak at the high-frequency limit remains, and the spectrum approaches that of the regular chain of the light atoms.

The main features mentioned above are roughly the same for both mass ratios 2 and 3. The effects of increase of mass ratio are merely to make the peaks more distinct and, for small f_L, to give rise to a few additional peaks just above the main band.

By a method which is a generalization of negative factor counting for secular matrices which are not tridiagonal, Martin (1961) calculated the spectra of chains with next-nearest-neighbour interactions, obtaining very similar results.

As will be explained in Chapter 4, there appear one or a few discrete *impurity frequencies* outside the main band of the regular lattice if it contains an island composed of a few impurity atoms lighter than the host atoms, occasionally with a few host atoms inserted between them. Dean found numerically that there is a remarkable correspondence between the frequencies of the distinct peaks in his spectrum and the impurity frequencies associated with several kinds of such islands of the light atoms with mass $m^{(0)}$ imbedded in the regular chain of heavy atoms with mass $m^{(1)}$. For example, the peaks *A*, *B*, *C*, *D*, *E*, *F*, and *G* in Figs. 2.5 and 2.6 correspond respectively to the islands L, LL, LLL or LHL, LHL, LL, LLL and LLLLL. (Here the letters L and H mean the light and heavy atoms, respectively.)

A disordered lattice can be considered as a lattice of heavy atoms containing many such islands randomly arranged. There are not only various kinds of islands, but a large number of islands of each kind. The above-mentioned correspondence suggests that even in a disordered chain each island contributes its own approximate impurity frequency to the spectrum of the whole system, notwithstanding the presence of other islands. The fact that such an independence of the individual island is preserved until the concentration of light atoms reaches about 0·6 is so striking that one is tempted to entertain a doubt as to whether such a suggestion corresponds to the truth. It will be shown in Chapter 5 that this, nevertheless, is actually the case.

Dean further showed that the strength of each peak is roughly proportional to the probability of occurrence of the corresponding island in the disordered lattice, and the above-mentioned concentration dependence of the spectrum may be explained by the change of the relative probabilities of different islands. This likewise affords a support for the above supposition.

The spectra of isotopically disordered two-dimensional diatomic lattices were calculated by Bacon and Dean (1962) and Dean and Bacon (1965). Figure 2.7 illustrates the spectra of the simple square lattice for various concentrations of light atoms. The mass ratio $m^{(1)}/m^{(0)}$ is 3, and $K = K'$. The broken curves for $f_L = 0{\cdot}1$ and 0·9 indicate the spectra of the regular monatomic lattice of heavy and light atoms respectively. The spectra of the honeycomb lattice, in which all the nearest-neighbour force constants are equal to a constant K and $m^{(1)}/m^{(0)} = 3$, are shown in Fig. 2.8.

The peak structure of these spectra and its dependence on the concentration is essentially the same, and can be explained in the same way, as in the case of one-dimensional chain. It seems, however, that the valleys begin to be filled up at a lower concentration than in the latter case. The correspondence between the peaks and the islands was also found. For example, the peaks *A* and *B* in Fig. 2.7 correspond to the islands L and L–L (any orientation), respectively, and the peak *C* in Fig. 2.8 to the island L.

Payton III and Visscher (1966) calculated the spectra of isotopically disordered simple cubic lattices with 6 × 6 × 25 atoms, in which there is no coupling between the displacements in different directions, and showed that the fine structure exists also in three-dimensional lattices.

Dean (1964) calculated the frequency spectrum of a chain with 7900 atoms in which the force constant varies randomly from atom to atom, while the mass is the same for all atoms (*glass-like chain*).

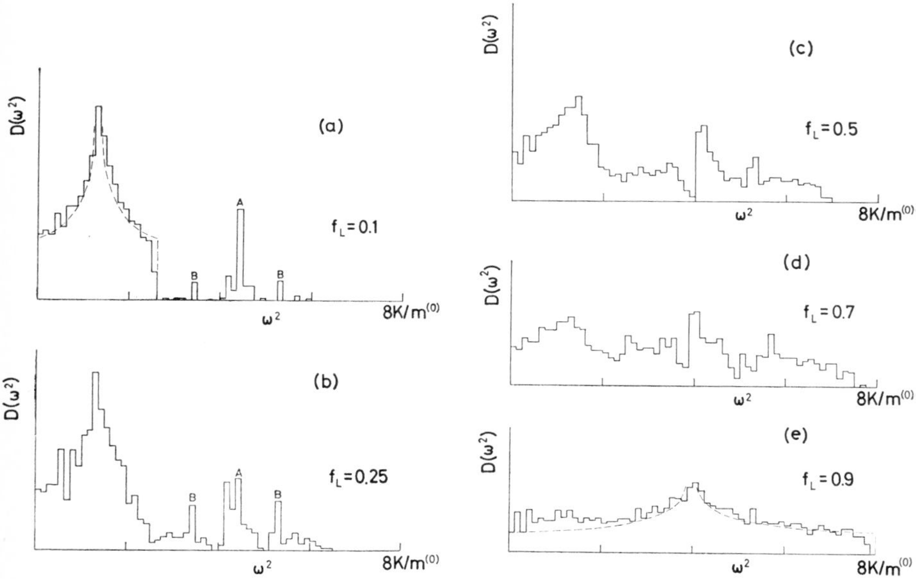

FIG. 2.7. Spectra for the isotopically disordered square lattice with mass ratio 3 and $K = K'$. The width and length of the sample lattices used are 16–18 and 50–64 respectively. The broken curves for $f_L = 0{\cdot}1$ and $0{\cdot}9$ indicate the spectra of the regular lattice of heavy and light atoms respectively. (After Dean and Bacon (1965); Crown copyright reserved.)

The distribution function of the force constant $K_{n;n+1}$ was chosen to be

$$P(K) = 1/(K_{max} - K_{min}), \quad \text{for} \quad K_{min} < K < K_{max},$$
$$= \quad 0, \quad \text{elsewhere}, \tag{2.4.1}$$

where $K_{n,n+1}$ was denoted simply by K. As shown in Fig. 2.9, the spectra are fairly smooth, in contrast to the above-mentioned cases. As the disorder increases, the high-frequency singularity of the regular chain becomes rounded and progressively less obvious. The small local fluctuations in the spectra were proved to be due to the finiteness of the chains used by comparing them with the spectra of other chains having the same statistical properties.

Dean also calculated the spectrum of a diatomic chain, in which the atoms with masses $m^{(0)}$ and $m^{(1)}(m^{(0)} < m^{(1)})$ are arranged alternatingly but the force constants are governed by the distribution (2.4.1). Typical results for the cases of mass ratios 1/2 and 1/4 are shown in Figs. 2.10 and 2.11. It is seen that the spectra are

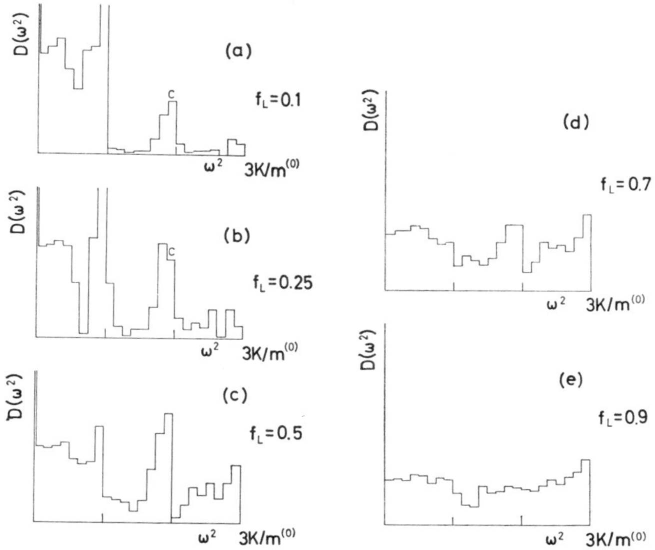

FIG. 2.8. Spectra for the disordered honeycomb lattice of the size 9×100, in which all the force constants are equal to a constant K and mass ratio is 3. (After Dean and Bacon (1965); Crown copyright reserved.)

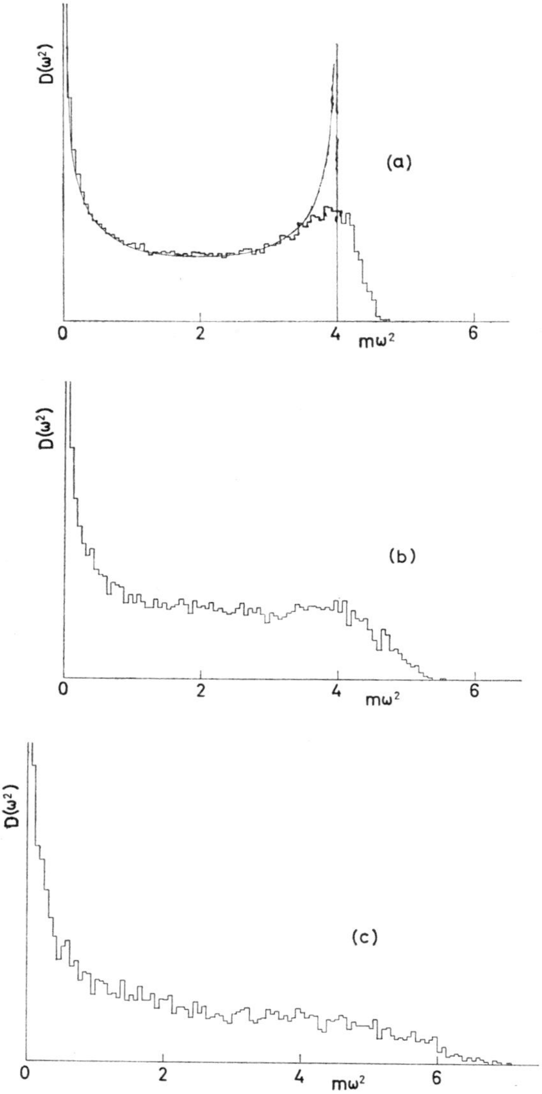

FIG. 2.9. Spectra for monatomic glass-like chains. (a) $K_{min} = 3/4$, $K_{max} = 5/4$; (b) $K_{min} = 1/2$, $K_{max} = 3/2$; (c) $K_{min} = 0$, $K_{max} = 2$. The smooth curve in (a) is the spectrum of the regular chain with $K = 1$. (After Dean (1964); Crown copyright reserved.)

smooth as in the monatomic case. The remarkable fact is that for small disorder the gap between two branches continues to exist. The limits of the gaps are approximately $m^{(0)}\omega^2 = 5/4$ and $3/2$ in Fig. 2.10a, $m^{(0)}\omega^2 = 5/8$ and $3/2$ in Fig. 2.11a and $m^{(0)}\omega^2 = 3/4$ and 1 in Fig. 2.11b.

The spectrum for the chain of infinite mass ratio can be calculated exactly. In this case the heavy masses become rigid walls, which

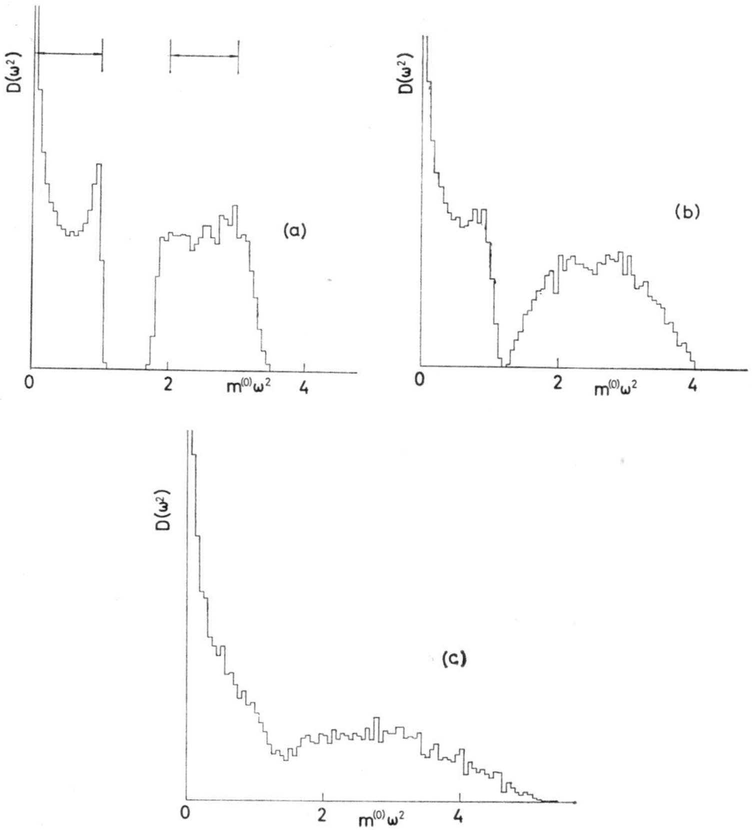

FIG. 2.10. Spectra for diatomic glass-like chains with mass ratio 2. (a) $K_{min} = 3/4$, $K_{max} = 5/4$; (b) $K_{min} = 1/2$, $K_{max} = 3/2$; (c) $K_{min} = 0$, $K_{max} = 2$. In (a) the regions of the acoustical and optical branches of the spectrum of the regular chain with $K = 1$ are indicated by horizontal arrows. (After Dean (1964); Crown copyright reserved.)

separate the light masses into $N/2$ independent single-atom systems. The frequency of vibration for the light atom at the nth site is given by

$$\omega^2 = (K_{n,n-1} + K_{n,n+1})/m^{(0)}. \tag{2.4.2}$$

It can easily be shown that, as $N \to \infty$, the distribution of ω^2 for light atoms becomes

$$\begin{aligned} D_L(\omega^2) &= m^{(0)}(m^{(0)}\omega^2 - 2K_{\min})/(K_{\max} - K_{\min})^2, \\ &\qquad 2K_{\min} \leqq m^{(0)}\omega^2 \leqq K_{\min} + K_{\max}, \\ &= m^{(0)}(2K_{\max} - m^{(0)}\omega^2)/(K_{\max} - K_{\min})^2, \\ &\qquad K_{\min} + K_{\max} \leqq m^{(0)}\omega^2 \leqq 2K_{\max}, \\ &= 0, \quad \text{elsewhere.} \end{aligned} \tag{2.4.3}$$

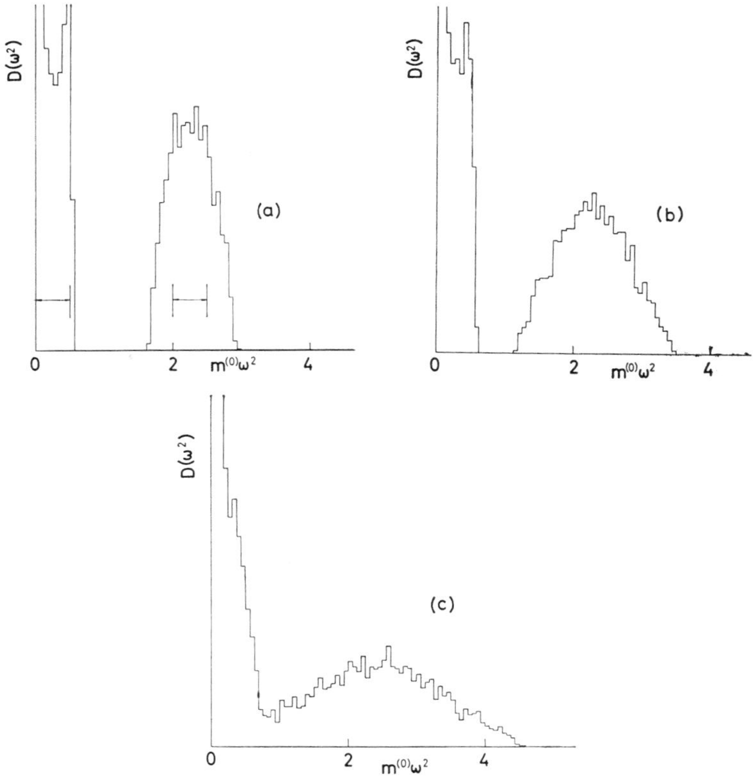

FIG. 2.11. Spectra for diatomic glass-like chains with mass ratio 4. (a) $K_{\min} = 3/4$, $K_{\max} = 5/4$; (b) $K_{\min} = 1/2$, $K_{\max} = 3/2$; (c) $K_{\min} = 0$, $K_{\max} = 2$. (After Dean (1964); Crown copyright reserved.)

Consequently, the total spectrum is

$$D(\omega^2) = \delta(\omega^2) + \tfrac{1}{2}D_L(\omega^2), \tag{2.4.4}$$

where the δ-function at the origin due to the heavy atoms contains half the frequencies of the system. The spectra in Figs. 2.11 b and 2.11 c are clearly approaching the triangular form (2.4.4).

2.5. Electronic Energy Spectra of Disordered Systems

The earliest of the numerical investigations of disordered systems was on the energy spectrum of an electron in a linear array of potentials. James and Ginzbarg (1953) and Landauer and Helland (1954) devised for the first time a node-counting technique and applied it to a random array of two kinds of square-well potentials. The sample chains they used contained only 150 potentials, so that only the band structure could be inferred from their calculations. Though owing to the insufficient length of the sample any definite conclusion could not be drawn, a suggestive result was obtained that there may be energy regions in which the levels are distributed in an irregular fashion without any well-defined band structure.

Landauer and Helland also studied the case of liquid metal in which the separation between successive potentials varies in a random manner. The result for the integrated spectrum in a band region will be referred to in § 7.4. They were pioneers also in the investigation of the problem of gap persistence in the liquid metal, but their results were too coarse compared to those obtained later by Makinson and Roberts and others, which will be mentioned below.

Agacy and Borland (1964) computed the energy spectrum of the electron moving in a random array of two different δ-potentials A and B with strengths $-V^{(A)}$ and $-V^{(B)}$ situated at $x = nl$ (l is a fixed length and n is integer), their probability of occurrence being f_A and $1 - f_A$ respectively (*disordered diatomic Kronig–Penney model*). In §§ 1.3 and 1.4 it was seen that such a system may be described by the state ratio u_n in the same way as those which are governed by a secular matrix $\mathbf{S}(\lambda)$. The equation for $u_n = \psi_n(0)/\psi_n'(0)$ can be obtained by substituting the matrix elements of the transfer matrix in (1.3.20) into (1.4.2). From such an equation Agacy and Borland derived a functional equation for the probability distribution of the phase δ_n in an infinitely long array. They solved it numerically and calculated the averaged spectrum.

The spectrum computed by Agacy and Borland for the case $lV^{(A)} = 4$, $lV^{(B)} = 0{\cdot}4$ and $f_A = 0{\cdot}5$ is shown in Fig. 2.12, where

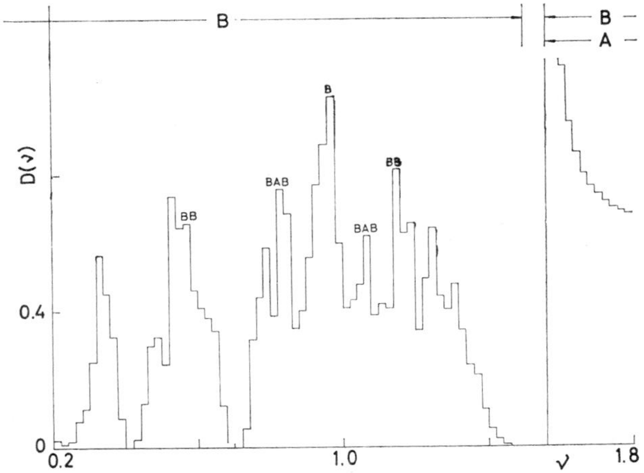

FIG. 2.12. The spectral density of the disordered diatomic Kronig–Penney model with $lV^{(A)} = 4$, $lV^{(B)} = 0{\cdot}4$ and $f_A = 0{\cdot}5$. $\nu \equiv 2k/V^{(A)}$. The energy bands of the regular chains of *A*- and *B*-potentials are indicated by horizontal lines with arrows. It is seen that the intervals $(0, \pi/2)$ and $(1{\cdot}504, \pi/2)$ correspond to the spectral gaps of the regular *A*- and *B*-chains respectively. The letters *B*, *BB* and *BAB* attached to the peaks indicate the corresponding islands. (After Agacy and Borland (1964); Crown copyright reserved.)

$\nu \equiv 2k/V^{(A)}$. Comparing this figure with Figs. 2.1–2.6, one sees that the spectrum is quite similar to the vibrational spectrum of an isotopically disordered chain. There are conspicuous peaks in the region from $\nu = 0$ to $\nu = 1{\cdot}504$, which lies in the energy gap of the regular array of *A*-potentials and in the energy band of the array of *B*-potentials. The energy region which was calculated out by Landauer and Helland to be without any well-defined band structure may be considered to correspond to such a peaky region. One can, moreover, associate the peaks with the electronic *im-*

purity levels characteristic of the islands of potentials B imbedded in the regular lattice of A, as indicated in the figure. It should further be noted that the interval $(1{\cdot}504, \pi/2)$, which is the spectral gap common to both regular systems, seems to be a gap also for the random array, and there exist three deep valleys at $\nu = 0{\cdot}23$, $\nu = 0{\cdot}41$, and $\nu = 0{\cdot}71$, at which the spectral density seems to vanish.

The electronic energy spectra of the disordered polymer models considered in § 1.2 were computed by Herzenberg and Modinos (1965) by using the method of Green function. They modified, however, the Green function by replacing the infinitesimal quantity ε by a finite one Γ. It can easily be seen that the effect of this replacement is to average the spectral density by the weighting function $\Gamma/\{\lambda_\mu - \lambda)^2 + \Gamma^2\}$, that is, to average the spectrum over an interval of order 2Γ. The model chosen by Herzenberg and Modinos was composed of two kinds of monomers A and B arranged randomly, the total number of monomer being 40. The quantities $\varepsilon_{n\alpha}$ and $P_{n\alpha 0}$ were chosen to be of the order of magnitude of corresponding quantities in the DNA helix. More precisely, the values $E_{n\alpha}^{(A)} = 4{\cdot}70$ and $E_{n\alpha}^{(B)} = 4{\cdot}81$ eV were used, corresponding to the averages of the prominent absorption maxima near 260 mμ in the adenine–thymine and guanine-cytosine pairs. The values used for $|P_{n\alpha 0}|^2$ were $|1{\cdot}03\ \text{A}e|^2$ and $|0{\cdot}73\ \text{A}e|^2$ for A- and B-monomers, respectively, where e is the electronic charge. The nearest-neighbour separation was taken to be $R = 3{\cdot}4$ A. All the calculations were made for the transition moments polarized perpendicular to the polymer axis, so that $g = +1$.

Figure 2.13a shows the spectrum obtained by taking $\Gamma = 0{\cdot}025$ eV, and Fig. 2.13b shows one of its short sections obtained by taking $\Gamma = 0{\cdot}005$ eV. Figures 2.13c, d, and e show three sections for $\Gamma = 0{\cdot}0025$ eV. Note that the regions of the energy bands of two regular chains composed of A and B (A- and B-chains) are (3·92, 5·48) and (4·42, 5·20) respectively. It is seen that there appear several peaks, especially outside the band of B-chain, so that the spectrum has a feature very similar to those obtained by Dean for isotopically disordered chains and by Agacy and Borland for the disordered diatomic Kronig–Penney model.

Eisenschitz and Dean (1957) calculated the spectrum of one-dimensional model of liquid potassium using the tight-binding approximation, mainly for the purpose of investigating whether

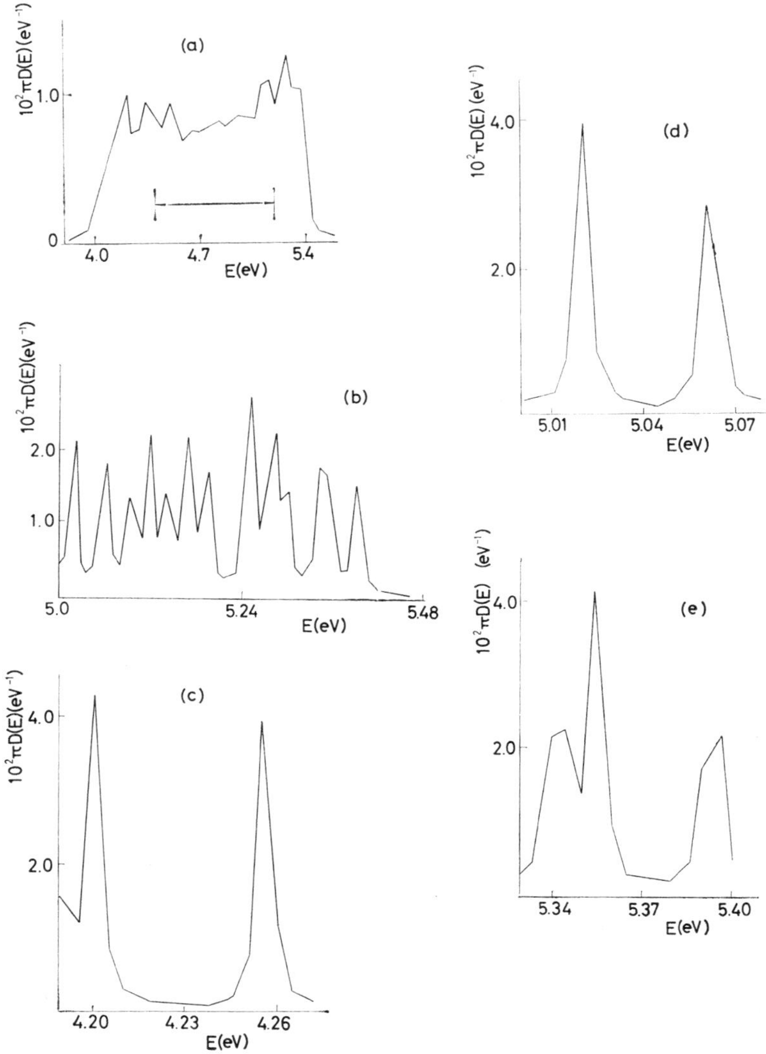

FIG. 2.13. Smoothed spectra for the polymer model of Herzenberg and Modinos. (a) $\Gamma = 0{\cdot}025$ eV. The complete region contains the whole band of *A*-chain. The horizontal arrow indicates the region of the band of *B*-chain. (b) A high-energy section obtained by taking $\Gamma = 0{\cdot}005$ eV. (c), (d), and (e) Three short sections obtained by taking $\Gamma = 0{\cdot}0025$ eV. (After Herzenberg and Modinos (1964).)

the spectral gaps persist or not when the regular crystals melt into the liquid state. They took the atomic orbital $\psi_n(x)$ from Hartree's tables of wave functions, and calculated the matrix elements of Hamiltonian

$$H = -\nabla^2 - \sum_n V(r - r_n), \tag{2.5.1}$$

where V is the potential of Hartree's self-consistent field by the procedure outlined in the end of § 1.2. They showed then that for the case of potassium the distribution of diagonal elements of H can be regarded directly as that of eigenvalues, and calculated the former for 4s and 5s bands. The result is shown in Fig. 2.14 in

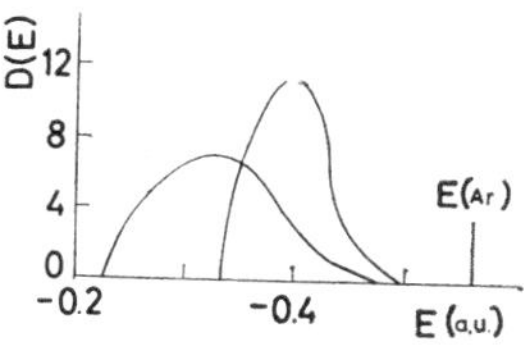

FIG. 2.14. Energy bands for liquid potassium. $E(\text{Ar}) = -0{\cdot}564$ (atomic unit) is the energy of the ground state of the isolated atom of argon. (After Eisenschitz and Dean (1957).)

which $E(Ar) = -0{\cdot}564$ a.u. indicates the energy of the ground 3s–3p state of the isolated atom of argon. Supposing that the 4s wave functions of argon and potassium do not differ markedly, they concluded that there is an energy gap between the 3s–3p and the 4s bands of argon, considering this as an explanation of the fact that the insulating property of the argon crystal is preserved at melting.

To investigate more firmly the problem of gap persistence, Makinson and Roberts (1960) and Roberts (1963) carried out a numerical calculation for a one-dimensional array of δ-potentials in which the interval l_n between neighbouring potentials varies randomly (*liquid-metal model*). The distribution of the spacing l_n was assumed to be the cut-off parabolic distribution

$$P(l_n) = \frac{3 \times 5^{1/2}}{20\sigma}\left\{1 - \frac{(l_n - 1)^2}{5\sigma^2}\right\}, \quad \text{for } 1 - 5^{1/2}\sigma \leq l_n \leq 1 + 5^{1/2}\sigma,$$
$$= 0, \text{ elsewhere}, \tag{2.5.2}$$

so that $\langle l_n \rangle$ is equal to unity and the standard deviation is σ. The computed integrated density spectra $M(k)$ near the first gap of the regular lattice with the spacing $\langle l_n \rangle$ are shown in Fig. 2.15, for the cases $V = 0{\cdot}25$ and $\sigma = 0{\cdot}02$ and $0{\cdot}05$. A gap seems to exist for $\sigma = 0{\cdot}02$, while it disappears for $\sigma = 0{\cdot}05$.

A similar calculation was made also by Borland and Bird (1964) by the method of functional equation. The distribution function for the spacing was assumed to be

$$
\begin{aligned}
P(l_n) &= 1/\Delta l, \quad |l_n - \langle l_n \rangle| \leqq \Delta l/2, \\
&= 0, \quad \text{elsewhere.}
\end{aligned}
\tag{2.5.3}
$$

In Fig. 2.16 the spectra for $\langle l_n \rangle\, V = 0{\cdot}8$ and $\sigma \equiv \Delta l/2 \times 3^{1/2} \langle l_n \rangle = 0$, $0{\cdot}04$, and $0{\cdot}08$ are illustrated. Again a gap seems to be remaining for $\sigma = 0{\cdot}04$ in the first gap of the regular system with $l_n = \langle l_n \rangle$. It disappears for $\sigma = 0{\cdot}08$. On the contrary, the second

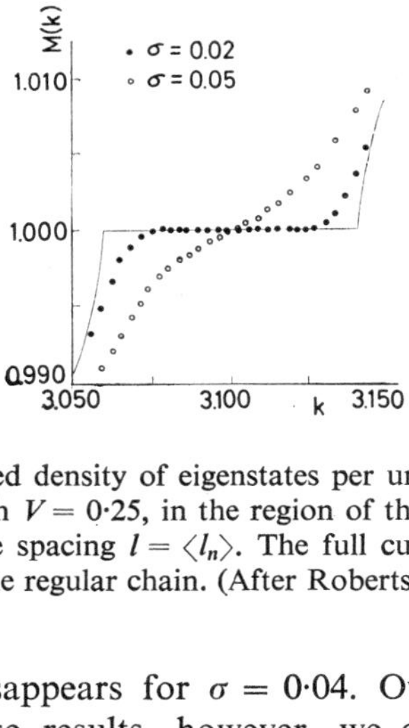

FIG. 2.15. The integrated density of eigenstates per unit length for the liquid-metal model with $V = 0{\cdot}25$, in the region of the first gap of the regular chain with the spacing $l = \langle l_n \rangle$. The full curve indicates the integrated density of the regular chain. (After Roberts (1963).)

gap completely disappears for $\sigma = 0{\cdot}04$. Owing to the numerical nature of these results, however, we cannot yet ascertain whether these gaps are true gaps or merely apparent ones, that is, whether the density is exactly zero there or merely extremely small. This problem will be considered in § 5.2. The complete solution has not yet been found, however.

Lax and Phillips (1958) considered a system of δ-potentials which are distributed at completely random positions, and calculated the impurity band of an electron moving in it, which originates from the bound state in the isolated δ-potential. This corresponds to the *effective-mass approximation* for the impurity band of the crystal containing impurity atoms. Frisch and Lloyd (1960) derived a

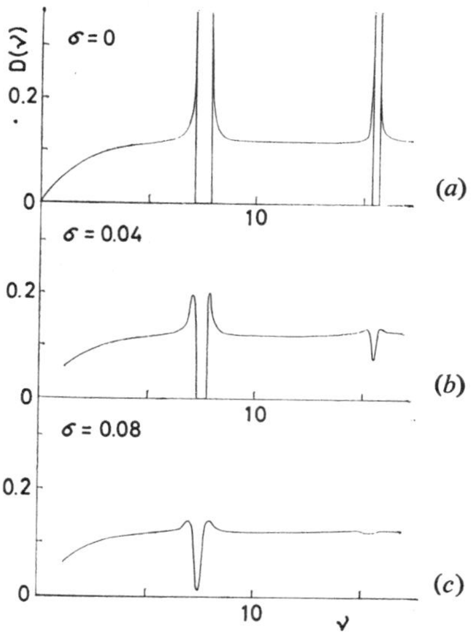

FIG. 2.16. The density of states for the liquid-metal model with $\langle l_n \rangle V = 0{\cdot}8$. $\nu \equiv 2k/V$. (a) $\sigma = 0$; (b) $\sigma = 0{\cdot}04$; (c) $\sigma = 0{\cdot}08$. (After Borland and Bird (1964); Crown copyright reserved.)

diffusion equation for the probability distribution of the vector variable $(\psi_n(0), \psi_n'(0))$ which undergoes a spatial stochastic change. In Fig. 2.17 the integrated spectra for some values of the concentration of potentials are shown, which were obtained by Lax and Phillips by direct node-counting and by Frisch and Lloyd by numerically solving their diffusion equation. It is seen that a narrow impurity band well separated from the main band appears for $c/\varkappa_0 = 0{\cdot}1$, whereas it begins to merge into the main band at $c/\varkappa_0 = 1$, where c is the concentration of potentials, and $\varkappa_0^2 \equiv -E_0$ the energy of the bound state of isolated potential. The explanation of the curves named optical model will be given in § 7.5.

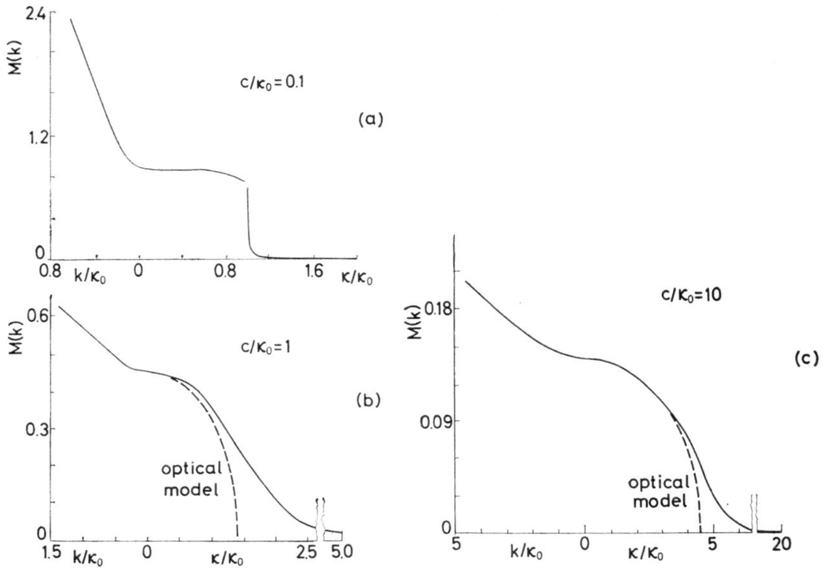

FIG. 2.17. The integrated density for the Kronig–Penney model with the spacings distributed completely at random. $k \equiv (+E)^{\frac{1}{2}}$, $\varkappa \equiv (-E)^{\frac{1}{2}}$, $\varkappa_0 \equiv (-E_0)^{\frac{1}{2}}$. (a) $c/\varkappa_0 = 0{\cdot}1$; (b) $c/\varkappa_0 = 1$; (c) $c/\varkappa_0 = 10$. (After Frisch and Lloyd (1960).)

2.6. Wave Form of Eigenmodes

The distribution of amplitude or intensity in each eigenmode of the disordered system is another important spectral property to be investigated. Rosenstock and McGill (1962) calculated the vibrational normal modes of some isotopic chains. Sample chains used by them were, however, so short ($N = 10 \sim 16$) that they could not draw any definite conclusion. Dean and Bacon (1963) carried out a computation for a particular sample of isotopically disordered diatomic chain

HLLLHLLHHHHHLLLLLLLHLHHLHHLHLHHLLLLLH
HLHLLLLHHHLHL (2.6.1)

which contains twenty-two heavy and twenty-eight light atoms. The mass ratio was taken to be 3. A part of their results is illustrated in Fig. 2.18. It is striking that the modes corresponding to high eigen-

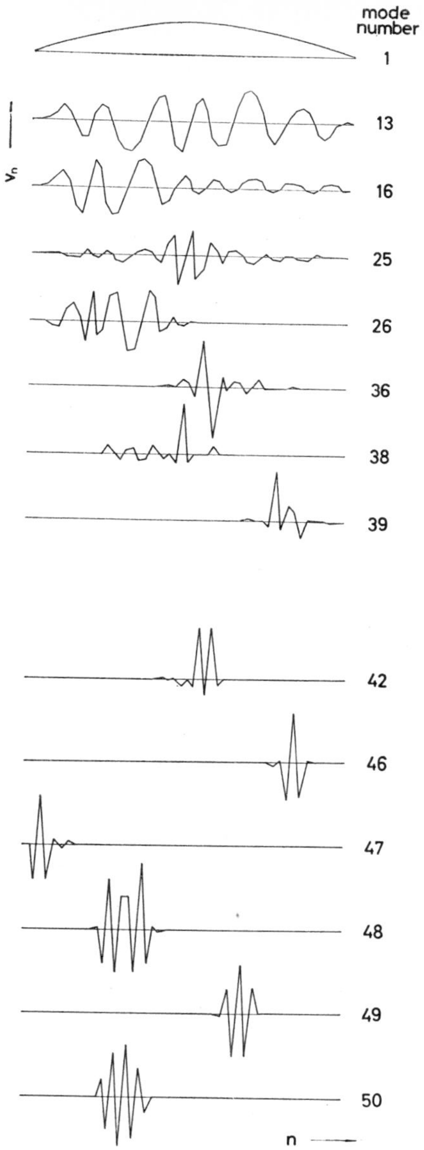

FIG. 2.18. Normal modes of the isotopically disordered chain with mass ratio 3. (After Dean and Bacon (1963); Crown copyright reserved.)

frequencies are strongly localized. The division between localization and non-localization is quite distinct for the system considered here, and occurs between the 25th and 26th eigenfrequencies. From Fig. 2.19, which shows the spectrum of the chain (2.6.1), it is seen that the value of the corresponding squared frequency is 3·0. Comparison with Fig. 2.4 shows that this frequency lies near the upper end of the smooth low-frequency part of the spectrum of the corresponding long chain.

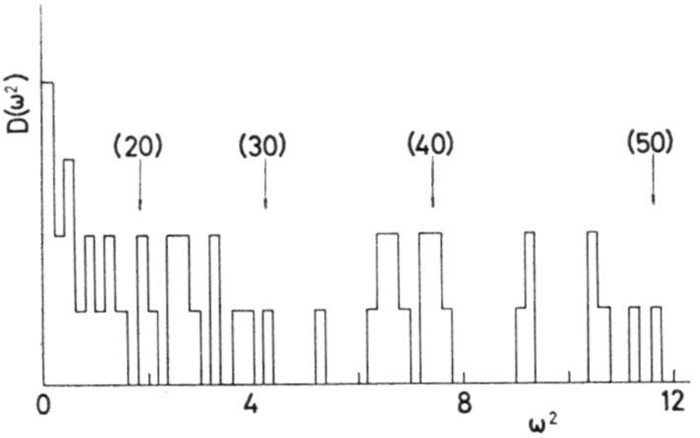

FIG. 2.19. Eigenfrequency spectrum of the same chain as that used for obtaining Fig. 2.18. The integers in brackets are the mode numbers. (After Dean and Bacon (1963); Crown copyright reserved.)

The position of localization of each mode with frequency above 4·0, which is the maximum frequency of the regular chain of heavy atoms, corresponds very closely to an island of light atoms. For example, in the 36th mode ($\omega^2 \sim 6{\cdot}7$), the localization occurs at

TABLE 2.1. *Correspondence between peak, island and the type of vibration at the position of localization*

Mode number	Squared-frequency	Peak	Island	Type of vibration
36	6·7	*C*	—LHL—	← →
38	7·2	*A*	—L—	
39	7·3	*A*	—L—	
42	7·6	*D*	—LHL—	← → ←
46	10·4	*F*	—LLL—	→ ← →
47	10·4	*F*	—LLL—	→ ← →
94	11·2	*G*	—LLLLL—	← ← → ← →

the island —LHL—. This is in harmony with the result mentioned in § 2.4 that the peak *C* with this frequency corresponds to the island —LHL—. In Table 2.1 such correspondences are shown.

Dean and Bacon showed, furthermore, that the type of vibration in the island at which the localization occurs closely corresponds to the vibration which the island would have at the corresponding frequency if it existed in isolation in a regular chain of heavy atoms. This is also shown in Table 2.1.

These facts again strongly support the supposition mentioned in § 2.4 that each island contributes its characteristic impurity frequency to the spectrum of the whole system, almost independently of other islands.

Payton III and Visscher (1966) calculated some vibrational eigenmodes of isotopically disordered square and simple cubic

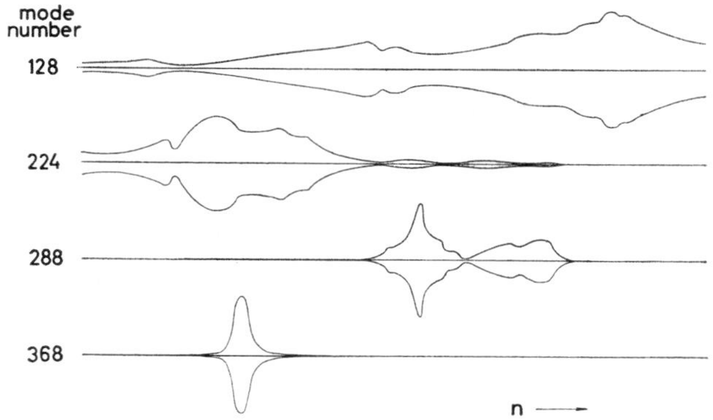

FIG. 2.20. Envelopes of the amplitude in the normal modes of a glass-like chain with 512 atoms, in which the distribution of force constants is given by (2.4.1), with $K_{min} = 3/4$ and $K_{max} = 5/4$. (After Dean (1964); Crown copyright reserved.)

lattices, and found that also in these cases the modes are strongly localized.

Similar localization occurs also in the glass-like chains. Figure 2.20 shows the envelopes of the typical normal modes of a monatomic glass-like chain computed by Dean (1964). It is seen that the localization becomes rapidly conspicuous with the increase of frequency.

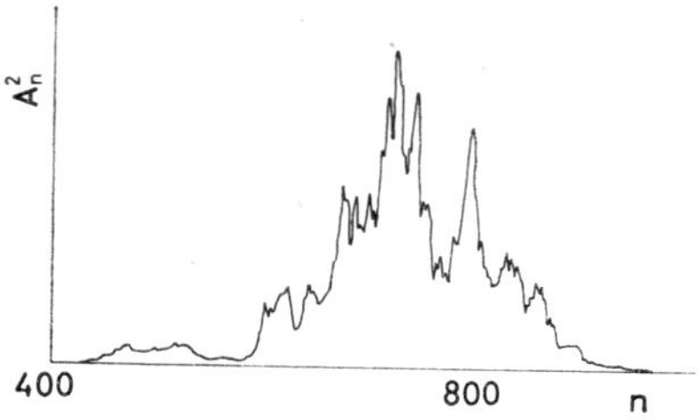

FIG. 2.21. The intensity distribution in the eigenfunction containing 19126 zeros of the liquid-metal model with $V\langle l_n\rangle = 10$ and $k/V = 6$. (After Borland (1963); Crown copyright reserved.)

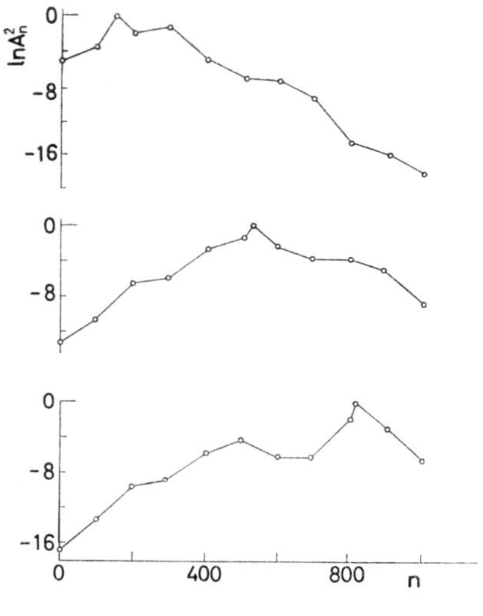

FIG. 2.22. The intensity distribution in three consecutive wave functions of the liquid-metal model with $V\langle l_n\rangle = 0{\cdot}1$ and $k/V = 2{\cdot}5$. (After Borland (1963); Crown copyright reserved.)

The situation remains the same also in the electronic spectrum. Borland (1963) considered a one-dimensional liquid-metal model in which the distance between neighbouring potentials l_n is distributed with the probability $P(l_n)$ given by

$$P(l_n) = \frac{1}{\langle l_n\rangle} \exp \frac{-l_n}{\langle l_n\rangle}. \tag{2.6.2}$$

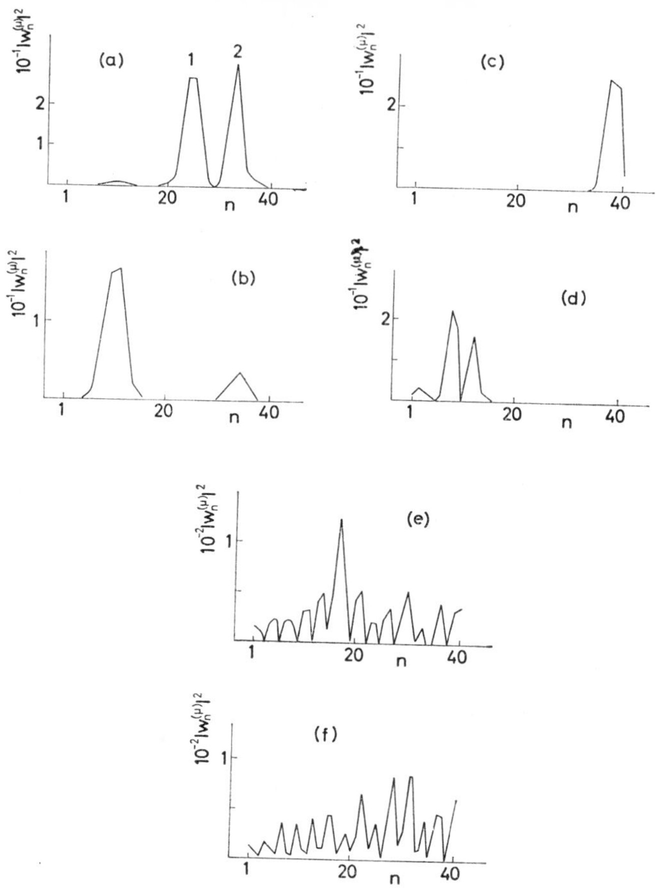

FIG. 2.23. The intensity distribution in the wave functions of the polymer model of Herzenberg and Modinos, calculated by taking $\Gamma = 0{\cdot}0025$ eV. The curves 1 and 2 in (a) correspond to the states at 5·395 and 5·355 eV respectively. Figures (b)–(f) show the states at 5·345, 4·255, 4·20, 5·06, and 5·02 eV respectively. (After Herzenberg and Modinos (1964).)

He calculated numerically the intensity of several wave functions, more precisely, the squared amplitude A_n^2 when the wave function ψ_n is represented as

$$\psi_n = \pm A_n \cos\{k(x - x_n) + \varphi_n\}, \tag{2.6.3}$$

where φ_n is the phase at x_n, and found that the functions are always clearly localized for large values of $V\langle l_n\rangle$, $-V$ being the strength of δ-potential. Figure 2.21 shows the intensity distribution in the eigenfunction containing 19126 zeros of a chain composed of 1000 potentials with parameters $V\langle l_n\rangle = 10$ and $k/V = 6$. It was also found that the wave functions are not all so clearly localized for small values of $V\langle l_n\rangle$, when the ratio k/V is kept equal to 6, but the reduction of k/V increases the degree of localization, so that clearly localized functions are obtained for $V\langle l_n\rangle = 0{\cdot}1$ and $k/V = 2{\cdot}5$. Figure 2.22 shows the plots of ln A_n^2 against n for three consecutive eigenfunctions calculated for these parameter values.

Such a localization of eigenfunctions was also demonstrated by Herzenberg and Modinos (1965). They calculated, by the method of Green function, the intensity distributions in some states of the polymer model mentioned in § 2.5. These are shown in Fig. 2.23. All curves were calculated by taking $\Gamma = 0{\cdot}0025$ eV. In Fig. 2.23a, two curves 1 and 2 are shown, which correspond to the states at 5·395 and 5·355 eV respectively. Figure 2.23b–f show the states at 5·345, 4·255, 4·20, 5·06, and 5·02 eV respectively. It is to be noted that the former five states, which lie outside the band of B-chain, are strongly localized, whereas the latter two, which lie inside that band, are not.

CHAPTER 3

Principles of Phase Theory

3.1. Phase and the Oscillation Theorems

In the first five sections of this chapter we consider the one-dimensional case, and discuss the properties of scalar state ratio and phase.

Consider the difference equation

$$(K_n' - \lambda M_n)\, v_n - K_n v_{n+1} - K_{n-1}\, v_{n-1} = 0, \qquad (3.1.1)$$

which is the one-dimensional version of the general equation (1.1.9) for the vibrational problem. Let us assume that $M_n > 0$ and $K_n > 0$, which is fulfilled in most systems of physical interest. The state ratio is now a scalar quantity defined by $u_n \equiv v_{n-1}/v_n$.

In the theory of difference equations, it is known that this quantity is a strictly increasing function of the energy parameter λ (Atkinson (1964)). The value of $u_n(\lambda)$ steadily increases with the increase of λ, tends to positive infinity as λ approaches one of the roots of the function $v_n(\lambda)$, and restarts to increase from the negative infinity after λ passes the root. Hence from what was said in § 1.4, the phase δ_n changes monotonically with λ, and passes a phase angle of α/γ when λ passes one of the roots of $v_n(\lambda)$, where α and γ are the coefficients of the Cayley-type transform (1.4.11). The direction of change of δ_n may be either positive or negative, depending upon the choice of the coefficients of Cayley-type transform. In the following we shall always choose them in such a manner that δ_n increases with λ.

For the boundary condition

$$v_0 = 0, \quad v_{N+1} = 0, \qquad (3.1.2)$$

N roots of the polynomial $v_{N+1}(\lambda)$ give the eigenvalues $\lambda_1, \lambda_2, \ldots, \lambda_N$ of (3.1.1). Now make out of the discrete series of values $v_1(\lambda)$, $v_2(\lambda), \ldots, v_{N+1}(\lambda)$ a continuous function $v_x(\lambda)$ of the variable x by connecting each neighbouring pair of values v_n and v_{n+1} by a straight line. For a fixed λ in the open interval $(\lambda_{r+1}, \lambda_{r+2})$, the function $v_x(\lambda)$ is known to change its sign just $r + 1$ times, and consequently there are $r + 1$ negative $u_n(\lambda)$. Therefore the number of negative u_n must be equal to the number of eigenvalues less than λ. This is just the negative-eigenvalue theorem stated in § 2.1 apart from the slightly different definition of the quantity u_n. The theorem may be regarded as a generalization of the above *oscillation theorem* to higher dimensional cases.

The theory of difference equations shows us further that all the zeros of $v_x(\lambda)$ move towards the left (direction of decreasing x) with the increase of λ. It is seen, therefore, from Fig. 3.1 that the function $u_n(\lambda)$ and $\delta_n(\lambda)$ can be extended to a unique continuous function of x, such that as x increases (for fixed λ) u_n passes the infinity and δ_n passes an angle of α/γ between two successive integral

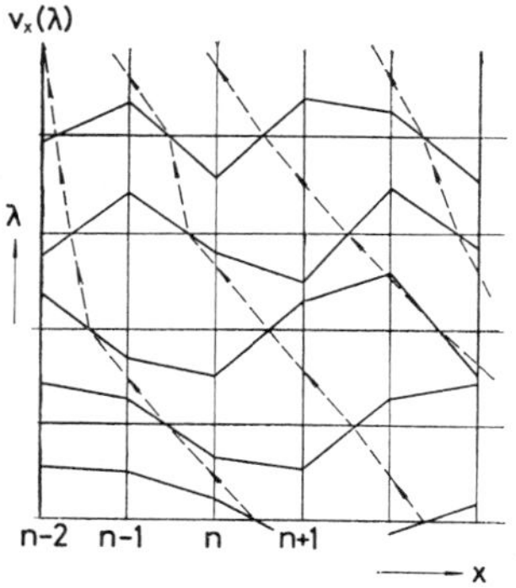

FIG. 3.1. The displacement of the zeros of $v_x(\lambda)$ with increase of the energy parameter λ.

values of x, whenever there is a zero of $v_x(\lambda)$ in this interval, while u_n never passes the infinity and δ_n never passes the angle of α/γ if there is no zero of $v_x(\lambda)$. Thus the number of eigenvalues less than λ is given by the number of points at which $\delta_x(\lambda)$ is equal to the angle of α/γ.

Also for the boundary conditions other than (3.2.1), the argument runs essentially in the same way. In any case, the number of eigenvalues less than λ is given by the number of points at which $\delta_x(\lambda)$ is equal to some fixed value (mod 2π). In other words, the number of eigenvalues less than λ is given by the integral part of the total phase change divided by 2π or of the total number of rotations of z_n.

The oscillation theorem similar to the above one is valid also for the differential equations of Sturm–Liouville type (Morse and Feshbach (1953)). For the Schrödinger equation it is known that the quantity $\psi(x)/\psi'(x)$ plays the role of $u_x(\lambda)$ in the above argument (James and Ginzbarg (1953), Landauer and Helland (1954)). It is seen from (1.3.18) and the definition of the state ratio given in § 1.4 that for the one-dimensional electronic systems such as considered in § 1.3 this quantity just coincides with the state ratio at the boundary of each cell. One may regard, therefore, the function $u_x(k) = \psi(x)/\psi'(x)$ as an extension of the state ratio $u_n(k)$ to the continuous range of x. Thus the number of points at which the phase $\delta_x(k)$ of $u_x(k)$ is equal to a fixed value, or the integral part of the total number of phase changes divided by 2π, gives the number of eigenvalues less than that corresponding to the value of k under consideration.

As has been mentioned in § 2.1, the above-stated property of the phase has been extensively used for numerical computation of the spectrum. However, the theoretical examination of the detailed behaviour of the phase has begun only recently. Schmidt (1957) and des Cloizeaux (1957) gave some discussions on the oscillative properties of the phase, but their purpose was to obtain a functional equation for the distribution of the negative factor or an expression for the characteristic function, so that the property of the phase itself was not fully utilized. It will be shown in what follows that, by combining the oscillative properties with other characteristic behaviours of the phase, which comes from the fact that it is transformed by successive linear transformations (1.4.2), one can make the phase one of the most powerful tools for investigating the spectral properties of disordered systems.

3.2. Complex Representations

As was already noted in § 1.4, it is often convenient to use, from the outset, the representation in which the complex state ratio z_n with modulus unity appears as the state ratio. It is clear that the matrix of the transformation **Y** from the original real representation to a complex one must be just the one associated with the Cayley-type transform (1.4.11), that is

$$\mathbf{Y} = \begin{pmatrix} \alpha & \beta \\ \gamma & \delta \end{pmatrix}, \tag{3.2.1}$$

where the elements α, β, γ and δ satisfy the condition (1.4.12) or (1.4.13).

Let the original real transfer matrix be

$$\mathbf{T} = \begin{pmatrix} p & q \\ r & s \end{pmatrix}. \tag{3.2.2}$$

Then the transformed matrix becomes

$$\mathbf{Q} = \mathbf{YTY}^{-1} = \delta(\alpha - \beta\gamma)^{-1} \begin{pmatrix} \alpha & \beta \\ \gamma & \delta \end{pmatrix} \begin{pmatrix} p & q \\ r & s \end{pmatrix} \begin{pmatrix} \delta & -\beta \\ -\gamma & \alpha \end{pmatrix}$$

$$= (\alpha\delta - \beta\gamma)^{-1} \begin{pmatrix} \alpha\delta p - \alpha\gamma q - \beta\delta r - \beta\gamma s & -\alpha\beta p + \alpha^2 q - \beta^2 r + \alpha\beta s \\ \gamma\delta p - \gamma^2 q + \delta^2 r - \gamma\delta s & -\beta\gamma p + \alpha\gamma q - \beta\delta r + \alpha\delta s \end{pmatrix}. \tag{3.2.3}$$

This matrix has the form

$$\mathbf{Q} = \begin{pmatrix} A & B \\ B^* & A^* \end{pmatrix}, \tag{3.2.4}$$

according to the condition (1.4.12) or (1.4.13). Conversely, the necessary condition for the transformed matrix to be of the form (3.2.4) is that **Y** induces a Cayley-type transform. A transfer matrix of the form (3.2.4), and also the representation in which the transfer matrix has such a form, are called *Cayley type*. We shall use the letter **Q** to denote the Cayley-type transfer matrices, and also in the general argument which does not depend on the type of transfer matrix.

Representations of Cayley type appear in various connections. Firstly, it can be shown that the transformation which diagonalizes a real unimodular matrix with complex eigenvalues is of Cayley type. As an example, consider a linear isotopic chain composed of atoms with masses $m^{(i)}(i = 0, 1, 2, \ldots$, in the order of increasing mass), and take the representation (1.3.11). Then the matrix at the atom with mass $m^{(i)}$ is given by

$$\mathbf{T}^{(i)} = \begin{pmatrix} 1 & 1 \\ -m^{(i)}\lambda/K & 1 - m^{(i)}\lambda/K \end{pmatrix}. \tag{3.2.5}$$

Let us choose a *standard mass* $m^{(b)}$ and define the wave-number parameter β by

$$\lambda = \omega^2 \equiv (4K/m^{(b)}) \sin^2 \beta. \tag{3.2.6}$$

The transfer matrix at the atom with the standard mass (the *basic atom*) is

$$\mathbf{T}^{(b)} = \begin{pmatrix} 1 & 1 \\ -m^{(b)}\lambda/K & 1 - m^{(b)}\lambda/K \end{pmatrix}. \tag{3.2.7}$$

For $\lambda < 2\{K/m^{(b)}\}^{1/2}$, the eigenvalues of $\mathbf{T}^{(b)}$ are given by $\exp(\pm 2i\beta)$, which are complex. On the contrary, for $\lambda > 2\{K/m^{(b)}\}^{\frac{1}{2}}$ the eigenvalues become real, and one should put $\beta = (\pi/2) + i\varepsilon$, ε being real.

In the representation in which $\mathbf{T}^{(b)}$ is diagonalized to become $\mathbf{Q}^{(b)} = \operatorname{diag}\{\exp(2i\beta), \exp(-2i\beta)\}$, the matrix $\mathbf{T}^{(i)}$ turns out to be

$$\mathbf{Q}^{(i)} = \begin{pmatrix} (1 + i\tan\gamma^{(i)})\exp(2i\beta) & i\tan\gamma^{(i)}\exp(-2i\beta) \\ -i\tan\gamma^{(i)}\exp(2i\beta) & (1 - i\tan\gamma^{(i)})\exp(-2i\beta) \end{pmatrix}, \tag{3.2.8}$$

where

$$\tan\gamma^{(i)} \equiv Q^{(i)}\tan\beta \equiv \{m^{(i)}/m^{(b)} - 1\}\tan\beta. \tag{3.2.9}$$

The matrix (3.2.8) is, in fact, of Cayley type. The matrix of transformation from the original representation (3.2.5) to the present one is

$$\mathbf{Y} = \frac{1}{2\sin 2\beta}\begin{pmatrix} i\exp(-2i\beta) & -i \\ -i\exp(2i\beta) & i \end{pmatrix}, \tag{3.2.10}$$

which is seen to satisfy the condition (1.4.12).

The basic atom may be chosen arbitrarily, and may not necessarily be one of the atoms actually present in the lattice under consideration. It is usually convenient to choose the atom which is lighter than or equal to the lightest atom composing the chain ($m^{(b)} \leqq m^{(0)}$). The parameter $Q^{(i)}$ which characterizes the mass ratio $m^{(i)}/m^{(b)}$ is called *mass parameter.*

As the second example, consider a disordered Kronig–Penney model. If we represent the wave function in the nth cell by $\psi_n = a_n \exp(ikx) + b_n \exp(-ikx)$, and take $(a_n, b_n)^T$ as the state vector $\mathbf{X}_n$, the transfer matrix which transfers $\mathbf{X}_n$ to $\mathbf{X}_{n+1}$ comes out to be a matrix of Cayley type:

$$\mathbf{Q}_n = \begin{pmatrix} \left\{1 + \dfrac{iV_n}{2k}\right\}\exp(ikl_n) & \dfrac{iV_n}{2k}\exp(-ikl_n) \\ \dfrac{-iV_n}{2k}\exp(ikl_n) & \left\{1 - \dfrac{iV_n}{2k}\right\}\exp(-ikl_n) \end{pmatrix}. \tag{3.2.11}$$

The transformation matrix from the representation (1.3.20) to the present one is

$$\mathbf{Y} = \tfrac{1}{2}\begin{pmatrix} k & -i \\ k & i \end{pmatrix}, \tag{3.2.12}$$

which induces a Cayley-type transformation, as is should be.

As has already been remarked in § 1.4, we shall denote the state ratio by z_n from the beginning in the Cayley-type representation. The phase δ_n is then simply the phase angle of the state ratio z_n.

It should be noted that the representations (3.2.8) and (3.2.11) have a defect that their elements become singular at $\beta = \pi/2$ and $k = 0$, respectively. This gives rise to some complications which should be treated with caution (see §§ 4.1 and 4.3). In case one wants to avoid such a difficulty, one should use the representations without any singularity, such as (1.3.8) and (1.3.20), or the representations which are obtained from them by Cayley-type transformations that are everywhere regular.

3.3. Classification of Linear Transformations

In the following three sections we investigate the general behaviours of the phase, which depend neither on the type of the original equation governing the system nor on the representation of the transfer matrix, but only on the fact that the variation of the phase from one lattice site to the next is described by a linear transformation (1.4.2), associated with the corresponding transfer matrix. Throughout these sections it is assumed that the transfer matrix is of Cayley type. It is evident from the foregoing considerations that this brings about no loss of generality.

At first consider the effect of a single transfer on the state ratio. Omit the suffix n and write

$$\mathbf{X}' = \mathbf{QX}. \tag{3.3.1}$$

Then the state ratio $z \equiv X^{(1)}/X^{(2)}$ is transferred to $z' = X'^{(1)}/X'^{(2)}$ by the linear transformation

$$z' = \frac{Az + B}{B^*z + A^*}. \tag{3.3.2}$$

In physical problems the matrix $\mathbf{Q}$ is always non-singular. It is known in the theory of functions of complex variable that for non-singular $\mathbf{Q}$ the linear transformation (3.3.2) induces a one-to-one mapping on the complex plane, which is conformal and transforms any circle to another circle.

If a point z is transformed by (3.3.2) into itself, i.e. if $z = z'$, it is called the *fixed point* of the transformation. It is obvious that in general there are two fixed points z_- and z_+ corresponding to two eigenvectors $\mathbf{X}_-$ and $\mathbf{X}_+$ of $\mathbf{Q}$: the ratios of components of $\mathbf{X}_-$ and $\mathbf{X}_+$ give z_- and z_+ respectively. The corresponding eigenvalues $\theta_\pm$ of $\mathbf{Q}$ are given by the roots of the equation

$$\theta^2 - 2(\mathrm{Re}\, A)\,\theta + \det \mathbf{Q} = 0. \tag{3.3.3}$$

Since $\det \mathbf{Q} = |A|^2 - |B|^2$ is always real, and by assumption is positive, we have only to distinguish the following three cases:

(1) $(\mathrm{Re}\, A)^2 > \det \mathbf{Q}$. In this case the eigenvalues $\theta_\pm$ are real and distinct. An arbitrary vector $\mathbf{X}$ can be represented as a linear combination of eigenvectors $\mathbf{X}_-$ and $\mathbf{X}_+$:

$$\mathbf{X} = c_-\mathbf{X}_- + c_+\mathbf{X}_+. \tag{3.3.4}$$

Operating $\mathbf{Q}$ on $\mathbf{X}$, we have

$$\mathbf{X}' = \mathbf{QX} = c_-\theta_-\mathbf{X}_- + c_+\theta_+\mathbf{X}_+. \tag{3.3.5}$$

Let us give the suffix $-$ to the eigenvalue with larger modulus and to the corresponding eigenvector and fixed point. Then the vector $\mathbf{X}'$ is obviously nearer to $\mathbf{X}_-$ than the initial vector $\mathbf{X}$ is, except when $\mathbf{X} = c_-\mathbf{X}_-$ or $\mathbf{X} = c_+\mathbf{X}_+$, in which case the direction of $\mathbf{X}$ is left invariant. Thus every vector except the eigenvector is "attracted" by $\mathbf{X}_-$ and "repulsed" by $\mathbf{X}_+$. Correspondingly, every point z on the complex plane, except the fixed points, approaches z_- and recedes from z_+. In other words, z_- and z_+ play the role of sink and source respectively of the flow field induced by (3.3.2) on the complex plane. According to the theory of linear transformation, the flow lines along which the points move under the transformation form a family of circles which pass through two fixed points, as is shown in Fig. 3.2. Circles which are orthogonal to the flow lines indicated by broken curves in the figure represent a series of successively mapped circles.

When the eigenvalues $\theta_\pm$ are real and distinct, the transfer matrix and the corresponding linear transformation are called *hyperbolic*. We shall call in this case the points z_- and z_+ *sink* and *source point*, and the vectors $\mathbf{X}_-$ and $\mathbf{X}_+$ *sink* and *source vector* respectively, considering their properties mentioned above. Further, we denote the phases of z_- and z_+ by δ_- and δ_+, and call them *sink* and *source phase* respectively. Occasionally the vectors $\mathbf{X}_\pm$, the points $z_\pm$, and the phases $\delta_\pm$ are collectively called *limit vectors, limit points*, and *limit phases*, respectively.

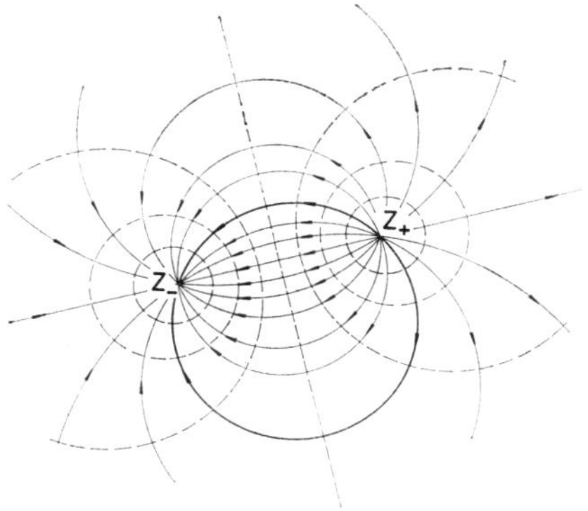

FIG. 3.2. Flow lines in the hyperbolic case. The arrows indicate the direction of the flow of the state ratio z. The broken curves show the family of successively transformed circles. The unit circle is depicted by a thick curve.

The limit points are easily seen to be given by

$$z_\pm = (\theta_\pm - A^*)/B^* = B/(\theta_\pm - A), \tag{3.3.6}$$

in terms of eigenvalues $\theta_\pm$. From (3.3.6) it is seen that $|z_\pm| = 1$. This is as it should be, since the state ratio must always be on the unit circle. We have only to examine the motion of points along this circle.

(2) $(\text{Re } A)^2 < \det \mathbf{Q}$. In this case the eigenvalues $\theta_\pm$ are distinct and complex conjugate with each other. Hence there can be no tendency for any point to approach or recede from a fixed point. In fact, the flow lines form a family of circles enclosing the fixed points. Every point rotates around one of the fixed points, neither approaching nor receding from it. The flow lines and a family of successively transformed circles are shown in Fig. 3.3. The points $z_\pm$ cannot obviously lie on the unit circle. In this case, the transfer matrix and the linear transformation induced by it are called *elliptic*.

(3) $(\text{Re } A)^2 = \det \mathbf{Q}$. In this intermediate case the eigenvalues $\theta_{\pm}$ coincide with each other, and consequently there exists only one fixed point $z_+ = z_-$. The flow lines are as shown in Fig. 3.4. It is seen that any point approaches this unique fixed point. The transformation is now called *parabolic*.

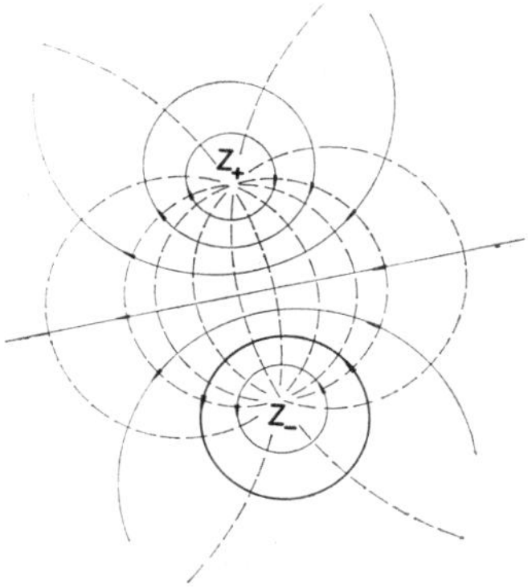

FIG. 3.3. Flow lines and family of successively transformed circles in the elliptic case. The unit circle is depicted by a thick curve.

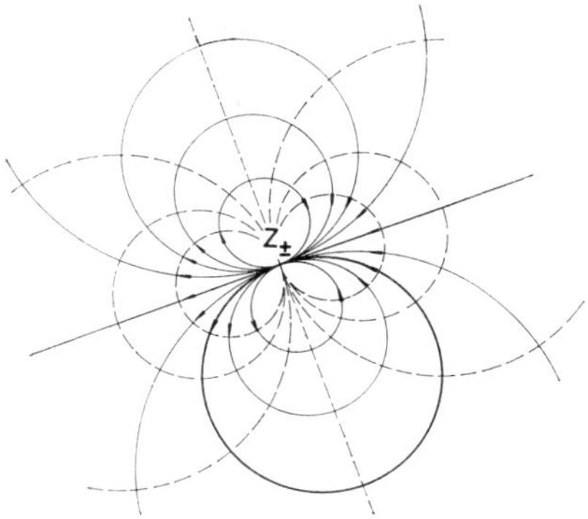

FIG. 3.4. Flow lines and family of successively transformed circles in the parabolic case.

It is clear that in any case the direction of the flow is reversed if we transfer the state vector towards the opposite direction. In the hyperbolic case, the sink and source points interchange their roles.

It should be remarked that (3.3.3), and consequently the above classification, does not depend on the representation of the transfer matrix.

3.4. Band Structure of Regular Systems

In this section we consider the regular system in which the same unit structure (*unit cell*) is repeated periodically, so that the system can be described by only one kind of transfer matrix **Q** corresponding to the unit cell. It was already remarked in § 1.3 that any unit cell can be described by a single transfer matrix, no matter how long and complex it may be.

In a regular system, the state ratio z_n is successively transformed by repeated operation of one and the same linear transformation induced by **Q** along the unit circle. Suppose now that the transformation is hyperbolic throughout an interval of the energy parameter λ (or β or k). Then for each value of λ in this interval, z_n is either fixed or approaches the sink point z_- from one of its sides. After a finite number of transfers, therefore, the amount of change of the phase during one transfer becomes nearly equal to an integral multiple of 2π, the multiplicity being determined by the nature of the unit cell. It is also a function of λ, but cannot change within such an interval of λ, since it must be continuous in λ. Thus the total amount of phase change over the whole system can increase with λ within this interval at most by a finite amount, even if the system is infinitely long. This means that there can exist at most only a finite number of eigenvalues in such a *hyperbolic interval* of λ. As the finite increase in the amount of phase change, if any, is completely determined during the finite number of initial transfers, the occurrence of these eigenvalues may be regarded as the end effect, so that they may be neglected in so far as one is concerned with the system which extends indefinitely towards both directions. At any rate the *density* of the spectrum must vanish in any hyperbolic interval. In other words, a hyperbolic interval must correspond to a *spectral gap* (or a *forbidden region*) of the regular system. The phase is, so to speak, *locked* at the sink phase, and its advance is prevented throughout the hyperbolic interval (Makinson and Roberts (1962)).

On the contrary, in an interval in which the transfer matrix is elliptic (*elliptic interval*), the phase is never locked but can increase freely with λ. More precisely, the amount of rotation of z during one transfer may increase steadily with λ everywhere in the lattice, so that the total amount of phase change is proportional to the extension of the system. This means that the spectral density does not in general vanish in an elliptic interval. In other words, any elliptic interval corresponds to a *spectral* (*energy* or *frequency*) *band*, or an *allowed region*, of the system. The value of λ at which the transfer matrix is parabolic corresponds always to a boundary between an elliptic interval and a hyperbolic one; it gives a boundary between a band and a gap.

As an example, consider a regular monatomic chain composed of atoms of mass m. If we choose the mass m as the standard mass ($m^{(b)} = m$), the transfer matrix is given by

$$\mathbf{Q} = \begin{pmatrix} \exp(2i\beta) & 0 \\ 0 & \exp(-2i\beta) \end{pmatrix} \tag{3.4.1}$$

in the Cayley-type representation (3.2.8). The matrix (3.4.1) is elliptic for $0 < \beta < \pi/2$, parabolic for $\beta = \pi/2$, and hyperbolic for $\beta = (\pi/2) + i\varepsilon$ (ε: real). This means that the frequency regions $\omega > 2(K/m)^{1/2}$ and $\omega < 2(K/m)^{1/2}$ are the spectral gap and the band, respectively, in harmony with the result obtained in § 1.6 that the maximum frequency of the regular monatomic chain is given by $2(K/m)^{1/2}$.

A less simple case is provided by the regular Kronig–Penney model. Let the strength of potential be $-V$ and the interval between potentials be l. In the representation (3.2.11), the transfer matrix is given by

$$\mathbf{Q} = \begin{pmatrix} \left\{1 + \dfrac{iV}{2k}\right\} \exp(ikl) & \dfrac{iV}{2k} \exp(-ikl) \\ -\dfrac{iV}{2k} \exp(ikl) & \left\{1 - \dfrac{iV}{2k}\right\} \exp(-ikl) \end{pmatrix}. \tag{3.4.2}$$

The condition for this matrix to be hyperbolic is

$$|A(k)| > 1, \tag{3.4.3}$$

where

$$A(k) \equiv \cos kl - (V/2k) \sin kl. \tag{3.4.4}$$

For negative energies one has to put $k = i\varkappa$, $\varkappa$ being real, and replace the trigonometric functions in the above formulae by the corresponding hyperbolic functions. The condition of hyperbolicity then becomes

$$|A(\varkappa)| > 1, \tag{3.4.5}$$

where

$$A(\varkappa) \equiv \cosh \varkappa l - (V/2\varkappa) \sinh \varkappa l. \tag{3.4.6}$$

Equations (3.4.3) and (3.4.5) are the well-known formulae of Kronig–Penney. In Fig. 3.5 the functions $A(k)$ and $A(\varkappa)$ are plotted,

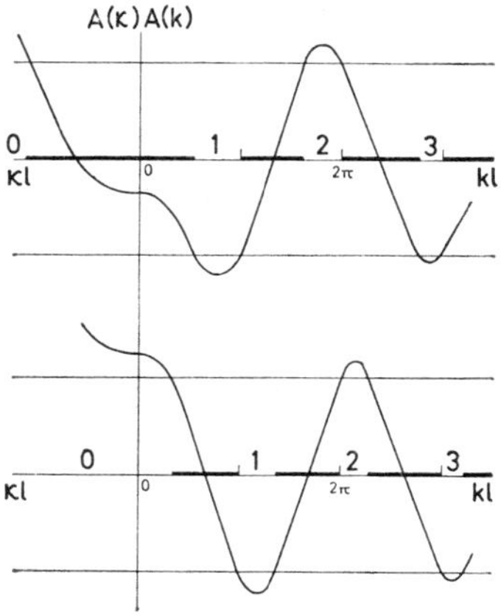

FIG. 3.5. A schematic diagram for the band structure of a regular Kronig–Penney model. The energy bands are indicated by thick lines, and the gaps are numbered as described in the text.

and the arrangement of the bands and gaps is illustrated. We shall call the gap below the lowest band which extends down to negative infinite energy the zeroth gap, and the other ones the first, second, third, ..., gaps, in the order of increasing energy. Further, we shall call the zeroth, second, ..., gaps *even*, and the first, third, ..., ones *odd*.

It is to be noted that for $V < 0$, the zeroth gap always includes a positive-energy part, while for $V > 0$ it lies wholly in the negative-energy region. The condition for the zeroth gap for $V > 0$ is

$$\tanh (\varkappa l/2) > V/2\varkappa. \tag{3.4.7}$$

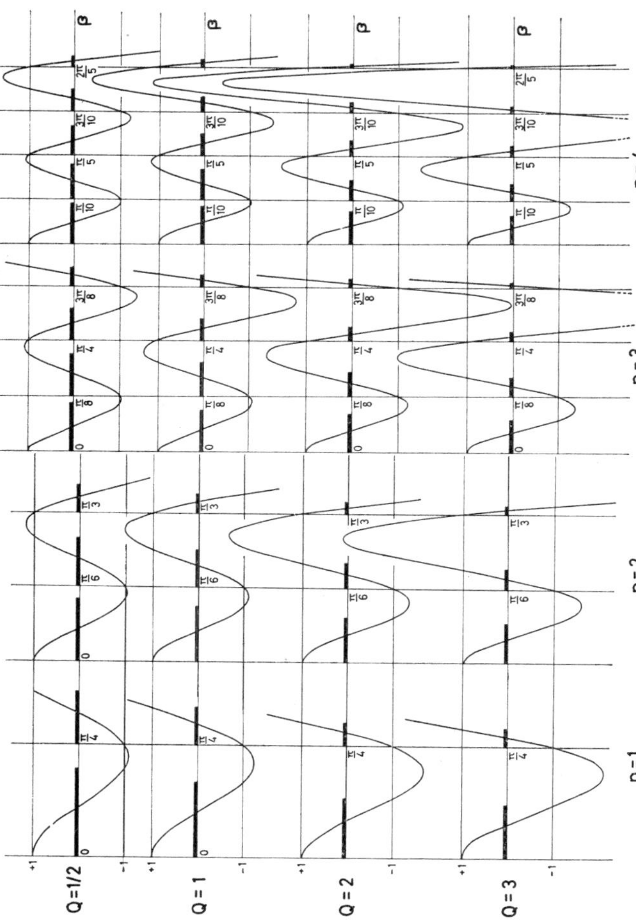

FIG. 3.6. The band structure of generalized diatomic chains with $p = 1, 2, 3, 4$ and $Q^{(1)} = m^{(1)}/m^{(0)} - 1 = \frac{1}{2}, 1, 2, 3$. The curves are plots of the left-hand side of (3.4.11) and the bands are depicted by thick lines. It is seen that the bands shrink while the gaps grow as the mass parameter $Q^{(1)}$ increases.

Further, for $V > 0$, the first gap extends to the negative-energy region if $Vl > 4$. Its extent is determined by the condition

$$\coth(\varkappa l/2) < V/2\varkappa . \tag{3.4.8}$$

Finally, let us investigate an example in which the unit cell has some structure. Consider an isotopic regular lattice with the unit cell composed of p light atoms of mass $m^{(0)}$ and one heavy atom of mass $m^{(1)}$ (*generalized diatomic lattice*; Maradudin and Weiss (1958) and Hori (1961)):

$$\underbrace{m^{(0)}m^{(0)}m^{(0)} \cdots m^{(0)}}_{p}m^{(1)} . \tag{3.4.9}$$

If we choose the light atom as the basic one, the transfer matrix $\mathbf{Q}^{(1p)}$ associated with the unit cell is given by, in the representation (3.2.8),

$$\mathbf{Q}^{(1p)} \equiv \mathbf{Q}^{(1)}\underbrace{\mathbf{Q}^{(0)}\mathbf{Q}^{(0)} \cdots \mathbf{Q}^{(0)}}_{p}$$

$$= \begin{pmatrix} \{1 + i\tan\gamma^{(1)}\}\exp\{2i(p+1)\beta\} & i\tan\gamma^{(1)}\exp\{-2i(p+1)\beta\} \\ -i\tan\gamma^{(1)}\exp\{2i(p+1)\beta\} & \{1 - i\tan\gamma^{(1)}\}\exp\{-2i(p+1)\beta\} \end{pmatrix}. \tag{3.4.10}$$

The condition for $\mathbf{Q}^{(1p)}$ to be hyperbolic is

$$\cos\{2(p+1)\beta\} - \tan\gamma^{(1)}\sin\{2(p+1)\beta\} > 1, \quad \text{or} \quad < -1 . \tag{3.4.11}$$

In Fig. 3.6 the function on the left-hand side is plotted and the band structure is illustrated for some values of p. As $m^{(1)} > m^{(0)}$, $\tan\gamma^{(1)}$ is always positive. It is seen that for any p-value, every band extends from $\beta = k\pi/2(p+1)$ to some upper value of β smaller than $(k+1)\pi/2(p+1)$, where k is an integer. The width of each band decreases as the mass ratio $m^{(1)}/m^{(0)}$ increases, but its bottom always remains fixed. It is particularly to be noted that the set of distinct values of $2\beta/\pi$ at these bottoms is just the set of all the rational numbers less than $\pi/2$ if one considers all the values of p from zero to infinity.

3.5. Spectral Gaps of Mixed Systems

The argument on the spectral gaps given in the preceding section can be generalized to the lattices which are described by two or more different transfer matrices. A regular lattice composed of several different atoms can be described either by a single matrix, as has been illustrated above, or by a set of different transfer matrices. On the contrary, a disordered lattice must necessarily be described by a number of matrices. The results of the present section are valid for any system which can be described by a set of different matrices, irrespective of whether it is regular or disordered. We shall call the system *mixed* when we are not concerned with its order.

Let a mixed system be described by a set of transfer matrices $\{\mathbf{Q}^{(i)}\}$, $i = 0, 1, 2, \ldots$ In regular systems the phase is transferred successively by a periodic series of $\mathbf{Q}^{(i)}$'s, while in disordered systems it is transferred by a random series of these matrices. We specify all the quantities pertaining to a particular transfer matrix, e.g. $\mathbf{Q}^{(j)}$ by the same index j. For example, the sink vector and sink phase of $\mathbf{Q}^{(j)}$ are denoted by $\mathbf{X}_-^{(j)}$ and $\delta_-^{(j)}$ respectively. Further, we call the regular system which is described by one of the members of the set, e.g. $\mathbf{Q}^{(k)}$, the kth *constituent* (*regular*) *system* of the given mixed system.

According to the result obtained in the previous section, one can readily say that, if in an interval of λ every transfer matrix belonging to the set $\{\mathbf{Q}^{(i)}\}$ is hyperbolic, that is if $(\mathrm{Re}\,A^{(i)})^2 > \det\mathbf{Q}^{(i)}$ for every i, this interval corresponds to a spectral gap for every constituent regular system. In such an interval, all the limit points $z_\pm^{(i)}$ lie on the unit circle. Let us call the interval spanned by the set of sink points $\{z_-^{(i)}\}$ *sink interval* and that spanned by $\{z_+^{(i)}\}$ *source interval*. Correspondingly, the intervals spanned by $\{\delta_-^{(i)}\}$ and $\{\delta_+^{(i)}\}$ are called *sink-phase* and *source-phase interval* respectively.

Suppose now that the sink-phase and source-phase intervals are disjoint. Then for any λ-value lying in the interval which corresponds to a gap for every constituent system, the phase eventually comes, during successive transformations, into the sink-phase interval, and once it has come in, it is trapped in it and can never escape therefrom. Thus there occurs again the locking of phase, so that its advance is prevented throughout the interval of λ under consideration. Thus we obtain the following theorem (Hori and Matsuda (1964), Hori (1966)):

THEOREM. *If* (*a*) *an interval of frequency or energy corresponds to a spectral gap for every constituent regular system, and* (*b*) *in that interval the sink- and source-phase intervals are disjoint, it gives a spectral gap of the mixed system described by the set* $\{\mathbf{Q}^{(i)}\}$.

If for the given set of matrices $\{\mathbf{Q}^{(i)}\}$ the condition (a) is a necessary consequence of the condition (b), we have the following proposition:

PROPOSITION. *If an interval of frequency or energy lies in a spectral gap for every constituent regular system, it gives a spectral gap of the mixed system.*

The set of matrices $\{\mathbf{Q}^{(i)}\}$ for which this proposition is true will be called *phase disjoint.*

The above proposition has an appearance very similar to the so-called Saxon–Hutner statement that if an energy interval corresponds to a spectral gap for every regular system which is composed of only one kind of the atoms in the mixed system, this interval corresponds to a gap of the mixed system. In fact, however, the above proposition is far more pregnant than the Saxon–Hutner statement. The reason is that the choice of the set $\{\mathbf{Q}^{(i)}\}$ describing the mixed system is to a large extent arbitrary. Owing to such a flexibility, it is rather exceptional that each constituent regular system coincides with a regular system composed of one kind of atoms. In this sense we call the proposition stated above a Saxon–Hutner-*type* proposition.

As an elementary example which illustrates the merit of the freedom in choosing the set $\{\mathbf{Q}^{(i)}\}$, consider an isotopically disordered diatomic lattice composed of atoms with masses $m^{(0)}$ and $m^{(1)}(m^{(0)} < m^{(1)})$, in which there appears no succession of light atoms containing more than $s - 1$ atoms. If we regard it as a mixed system composed of two kinds of atoms and adopt as the set $\{\mathbf{Q}^{(i)}\}$ the set $\{\mathbf{Q}^{(0)}, \mathbf{Q}^{(1)}\}$, for example, the above proposition means trivially that there is no eigenfrequency larger than the maximum frequency of the regular lattice of light atoms, since the interval in which both $\mathbf{Q}^{(0)}$ and $\mathbf{Q}^{(1)}$ are hyperbolic is $\lambda > 4K/m^{(0)}$. We may, however, choose the set $\{\mathbf{Q}^{(1p)}\}$, $p = 0, 1, 2, \ldots, s - 1$, given by (3.4.10), as the set of matrices describing the system. Then the generalized diatomic lattices with p-values from zero to $s - 1$ just become the constituent lattices. It can readily be seen from Fig. 3.6 and the remark given at the end of § 3.4 that, for a sufficiently large mass parameter $Q^{(1)}$, there appear, just below one or some of the bands of constituent lattices, narrow β-intervals, each of which corresponds to a gap for every constituent lattice.

Thus it may well be expected that, if we choose the set $\{\mathbf{Q}^{(i)}\}$ in an appropriate manner, our proposition may lead to a quite unexpected result which is beyond the scope of the ordinary Saxon–Hutner statement. In fact, by making the above argument more precise, we can reach an important conclusion about the structure of the spectrum of a disordered isotopic chain. The detailed account is, however, postponed to Chapter 5, where we shall apply the Saxon–Hutner-type proposition to several disordered systems, showing thereby that the condition of phase disjointness is satisfied in these cases.

3.6. Principles of Continued-fraction Theory

Matsuda (1962a) and Matsuda and Okada (1965) found that the continued-fraction expansion (1.4.6) of the phase is very useful for investigating the problem of spectral gaps of disordered systems. In some cases we can reach the same result in much simpler way by using the continued-fraction theory than through the theorem of Hori–Matsuda. Its most important merit lies, however, in that it can be generalized to higher-dimensional systems without much difficulty, in contrast to the arguments described in the preceding sections of this chapter. In the present section we derive the general continued-fraction expansion, and state the fundamental principles of the theory of Matsuda and Okada.

Let us write the general matrix-difference equation (1.3.2) in the vector-matrix form:

$$\begin{pmatrix} \mathbf{V}_{n-1} \\ \mathbf{V}_n \end{pmatrix} = \begin{pmatrix} -\mathbf{D}_n^{T-1}\mathbf{C}_n & -\mathbf{D}_n^{T-1}\mathbf{D}_{n+1} \\ \mathbf{I} & \mathbf{0} \end{pmatrix} \begin{pmatrix} \mathbf{V}_n \\ \mathbf{V}_{n+1} \end{pmatrix} = \mathbf{Q}_n \begin{pmatrix} \mathbf{V}_n \\ \mathbf{V}_{n+1} \end{pmatrix}, \tag{3.6.1}$$

which is different from (1.3.4) only in that the direction of transfer is reversed. We have then

$$\begin{pmatrix} \mathbf{V}_n \\ \mathbf{V}_{n+1} \end{pmatrix} = \mathbf{Q}_{n+1;\,k} \begin{pmatrix} \mathbf{V}_k \\ \mathbf{V}_{k+1} \end{pmatrix}, \tag{3.6.2}$$

where

$$\mathbf{Q}_{nk;} \equiv \mathbf{Q}_n \mathbf{Q}_{n+1} \dots \mathbf{Q}_k. \tag{3.6.3}$$

If we impose the fixed boundary condition at $n = 1$ and N, we must have

$$\mathbf{V}_0 \mathbf{z} = \mathbf{0}, \tag{3.6.4}$$

and consequently from (3.6.2)

$$\mathbf{Q}_{1;N}^{(11)}\mathbf{z} = \mathbf{0}, \tag{3.6.5}$$

according to (1.5.3) and (1.5.4). The condition for (3.6.5) to have a non-trivial solution is

$$\det \mathbf{Q}_{1;N}^{(11)} = 0. \tag{3.6.6}$$

Introducing as in § 1.4 the state ratio matrix $\mathbf{U}_n \equiv \mathbf{V}_{n-1}\mathbf{V}_n^{-1}$, we find, by using (3.6.2),

$$\mathbf{U}_n = \mathbf{V}_{n-1}\mathbf{V}_n^{-1} = \mathbf{Q}_{n;N}^{(11)}\mathbf{Q}_{n+1;N}^{(11)-1}. \tag{3.6.7}$$

The condition (3.6.6) is therefore satisfied if either

$$\det \mathbf{U}_1 = 0 \tag{3.6.8}$$

or

$$\det \mathbf{Q}_{2;N}^{(11)} = 0. \tag{3.6.9}$$

The latter condition is, however, compatible with (3.6.6) only for a particular choice of $\mathbf{D}_2$, as is seen from the equation

$$\mathbf{Q}_{1;N}^{(11)} = \mathbf{D}_1^{T-1}\mathbf{C}_1\mathbf{Q}_{2;N}^{(11)} - \mathbf{D}_1^{T-1}\mathbf{D}_2\mathbf{Q}_{2;N}^{(21)}. \tag{3.6.10}$$

It fails to hold, therefore, if the matrix $\mathbf{D}_2$ is changed in any way. Since we are interested in the property which does not depend on the boundary condition, we may consider that the condition (3.6.6) is equivalent to (3.6.8).

As in § 1.4, the matrix $\mathbf{U}_{n(k)}$ is related with $\mathbf{U}_{n(k+1)}$ by a linear transformation

$$\mathbf{U}_{n(k)} = \{\boldsymbol{\alpha}_k\mathbf{U}_{n(k+1)} + \boldsymbol{\beta}_k\}\{\boldsymbol{\gamma}_k\mathbf{U}_{n(k+1)} + \boldsymbol{\delta}_k\}^{-1}, \tag{3.6.11}$$

where

$$\begin{pmatrix} \boldsymbol{\alpha}_k & \boldsymbol{\beta}_k \\ \boldsymbol{\gamma}_k & \boldsymbol{\delta}_k \end{pmatrix} \equiv \mathbf{Q}_{n(k)}\mathbf{Q}_{n(k)+1} \cdots \mathbf{Q}_{n(k+1)-1},$$

$$1 = n(0) < n(1) < n(2) < \ldots < n(l+1) = N+1. \tag{3.6.12}$$

From (3.6.11) we obtain the general continued-fraction expansion

$$\mathbf{U}_1 = \mathbf{b}_0 + \cfrac{\mathbf{a}_1}{\mathbf{b}_1 + \cfrac{\mathbf{a}_2}{\mathbf{b}_2 + \cfrac{\mathbf{a}_3}{\mathbf{b}_3 + \ddots + \cfrac{\mathbf{a}_l}{\mathbf{b}_l}}}}, \tag{3.6.13}$$

where

$$\mathbf{b}_0 \equiv \boldsymbol{\alpha}_0/\boldsymbol{\gamma}_0, \quad \mathbf{b}_k \equiv \boldsymbol{\delta}_{k-1} + \boldsymbol{\gamma}_{k-1}\boldsymbol{\alpha}_k/\boldsymbol{\gamma}_k, \quad k = 1, 2, \ldots, l,$$
$$\mathbf{a}_1 \equiv \boldsymbol{\beta}_0 - \boldsymbol{\alpha}_0\boldsymbol{\gamma}_0^{-1}\boldsymbol{\delta}_0, \; \mathbf{a}_k \equiv \boldsymbol{\gamma}_{k-2}(\boldsymbol{\beta}_{k-1} - \boldsymbol{\alpha}_{k-1}\boldsymbol{\gamma}_{k-1}^{-1}\boldsymbol{\delta}_{k-1}),$$
$$k = 2, 3, \ldots, l. \tag{3.6.14}$$

Here we denoted, for example, the matrix $\boldsymbol{\alpha}_0\boldsymbol{\gamma}_0^{-1}$ by $\boldsymbol{\alpha}_0/\boldsymbol{\gamma}_0$. Equation (3.6.13) can also be rewritten as

$$\mathbf{U}_1 = \{\mathbf{I} + \mathbf{U}(1, l)\}\, \mathbf{b}_0,$$

$$\mathbf{U}(1, l) = \cfrac{\mathbf{a}_1\mathbf{b}_1^{-1}}{\mathbf{I} + \cfrac{\mathbf{a}_2\mathbf{b}_2^{-1}}{\mathbf{I} + \ddots + \cfrac{\mathbf{a}_{l-1}\mathbf{b}_{l-1}^{-1}}{\mathbf{I} + \mathbf{a}_l\mathbf{b}_l^{-1}\mathbf{b}_{l-1}^{-1}}\,\mathbf{b}_{l-2}^{-1}}\,\mathbf{b}_1^{-1}}\,\mathbf{b}_0^{-1}. \tag{3.6.15}$$

Since we are concerned with the property which is independent of the boundary condition, we may assume

$$\det \mathbf{b}_0 \neq 0. \tag{3.6.16}$$

Then it is seen from (3.6.8) that, if for a value of λ in the interval (λ_1, λ_2) the continued fraction $\mathbf{U}(1, l)$ converges in the limit $l \to \infty$ to the value

$$\mathbf{U} = \lim_{l \to \infty} \mathbf{U}(1, l), \tag{3.6.17}$$

which satisfies

$$\det(\mathbf{I} + \mathbf{U}) \neq 0, \tag{3.6.18}$$

this interval lies in a spectral gap.

That the quantity $\mathbf{U}(1, l)$ converges to the value such that (3.6.8) holds means that the condition at the remote boundary of the system does not affect the eigenvalue. Therefore we can consider that if for some λ $\mathbf{U}(1, l)$ converges to the value satisfying (3.6.8), this value of λ corresponds to an impurity frequency or level, since, as will be shown in Chapter 4, such a frequency or level does not depend on the boundary condition.

If $\mathbf{U}(1, l)$ converges uniformly in a region (λ_1, λ_2), the limit $\mathbf{U}$ is a continuous function of λ. This would mean that there are only isolated points at which $\det(\mathbf{I} + \mathbf{U}) = 0$, for if this determinant accidentally vanishes throughout a certain subinterval, it is possible to make it non-vanishing by merely changing the transfer matrices in the boundary region. Thus we may state that, as far as the spectral density is concerned, the interval in which $\mathbf{U}(1, l)$ converges uniformly in λ corresponds to a spectral gap, though it may contain isolated impurity levels which do not contribute to the spectral density. On the contrary, if the convergence is not uniform and (3.6.18) is not always satisfied in the interval, there may be many impurity levels which contribute to the density, so that we cannot assert that the interval belongs to a spectral gap. If, finally, $\mathbf{U}(1, l)$ diverges in some interval of λ, it is clear that this interval cannot correspond to any gap.

Thus in order to investigate the gap problem, we have only to examine the convergence property of the continued fraction (3.6.15). It is important that one may choose the series of site number $n(k)$, $k = 1, 2, \ldots, l$, in an entirely arbitrary manner. This corresponds to the freedom of choice of the set of transfer-matrices $\{\mathbf{Q}^{(i)}\}$ describing the system, which has already been shown to be one of the most important sources of power of the phase theory.

In the one-dimensional case, all the quantities appearing in the continued-fraction expansion (3.6.15) reduce to scalars, which are to be denoted by the corresponding light-faced letters. In such a case we have a useful theorem due to Worpitzky (Wall (1948)):

THEOREM. *If, in a continued fraction*

$$w_n = \cfrac{\varrho_1}{1 + \cfrac{\varrho_2}{1 + \cfrac{\varrho_3}{1 + \cfrac{\varrho_4}{\ddots + \cfrac{\varrho_{n-1}}{1 + \varrho_n}}}}}, \tag{3.6.19}$$

where ϱ_k's are functions of any variable over a domain D, the inequalities

$$|\varrho_k| \leqq 1/4, \quad k = 1, 2, \ldots \tag{3.6.20}$$

are satisfied, then

(*a*) *w_n converges uniformly over D;*

(*b*)

$$\left| \frac{1}{1 + w_n} - \frac{4}{3} \right| \leqq \frac{2}{3}, \tag{3.6.21}$$

and

(*c*) *the constant* 1/4 *is the best constant that can be used in* (3.6.20), *and* (3.6.21) *is the best domain of values of w_n.*

The use of this theorem will be illustrated in § 5.5.

CHAPTER 4

Description of Impurity Modes by the Phase Theory

4.1. Rate of Convergence of the Phase

Let us again consider how the phase changes during the repeated transformations by a single transfer matrix. It was shown in § 3.3 that when the matrix is hyperbolic, every phase except δ_- and δ_+ approaches or converges to the sink phase δ_-. Suppose now that every phase lying outside the interval $(\delta_+ - \Delta_+, \delta_+ + \Delta_+)$ with a specified small width $2\Delta_+$ comes into the interval $(\delta_- - \Delta_-, \delta_- + \Delta_-)$ with another specified width $2\Delta_-$ after at most r transfers. The integer r, which is a function of Δ_+, Δ_-, and λ, is called the *rate of convergence* associated with the given transfer matrix, and denoted by $r(\Delta_+, \Delta_-, \lambda)$.

Hori and Fukushima (1963) calculated numerically the rate of convergence for the case of a monatomic chain with the atomic mass $m^{(1)}$, using the transfer matrix (3.2.8) with $i = 1$. They took $Q^{(1)} \equiv (m^{(1)}/m^{(b)}) - 1 = 1$, $\beta = 0{\cdot}955$ rad, and calculated the phase variations from various initial phases. The chosen value of β corresponds to the impurity frequency of an atom of mass $m^{(b)}$ imbedded in a regular lattice composed of atoms with mass $m^{(1)}$. The phase φ used by Hori and Fukushima is somewhat different from the present one δ. The relation between them is

$$\tan \varphi = -\cos(\gamma^{(1)} + \delta + 4\beta)/\{1 - \sin(\gamma^{(1)} + \delta + 4\beta)\}, \quad (4.1.1)$$

from which it is seen that an interval of δ with the width 2π is monotonically mapped into the interval $-\pi/2 < \varphi \leqq \pi/2$.

From Fig. 4.1, which illustrates the results obtained by Hori and Fukushima, it is immediately seen that in the forward transfer the convergence towards the sink phase is surprisingly rapid, and the source phase is extremely unstable. For example, they obtained the following values for the rate of convergence:

$$r(0{\cdot}032,\ 0{\cdot}01,\ 0{\cdot}955) = 4,$$
$$r(0{\cdot}032,\ 0{\cdot}001,\ 0{\cdot}955) = 5,$$
$$r(0{\cdot}032,\ 0{\cdot}0001,\ 0{\cdot}955) = 6. \tag{4.1.2}$$

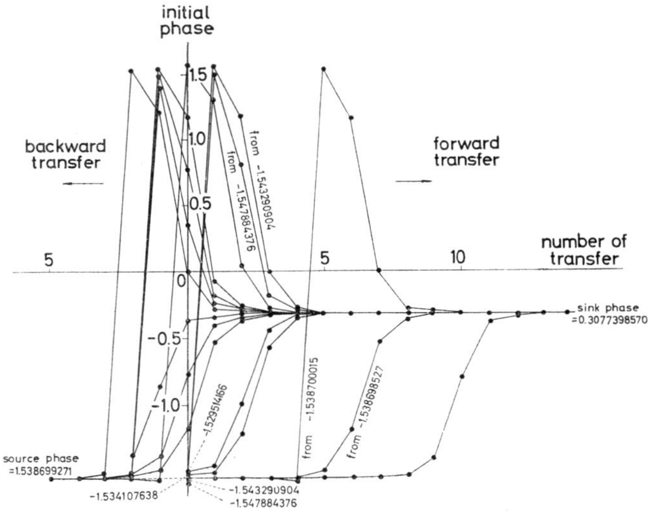

FIG. 4.1. Convergence of the phase towards the sink and source phases in the forward and backward transfers, respectively. The unit of phase angle is the radian.

(The unit of angle is the radian.) Even the phase $-1{\cdot}538699271$, which has been calculated as the source phase, converges to the sink phase after only thirteen, fourteen, and fifteen transfers, with the errors 0·01, 0·001, and 0·0001 respectively. The error of the computation has been at most 10^{-8}. Such a slight deviation from the true source phase already causes a strikingly rapid approach to the sink phase. The situation remains exactly the same also in the backward transfer except that the role of sink and source phases is interchanged. The phase converges rapidly to the source phase, and the sink phase becomes extremely unstable. This means that, by choosing a phase within a very narrow interval around the sink phase, one may obtain almost arbitrary phase by transferring it towards the reverse direction only a few times. Actually, by choosing an initial phase within the interval of widths 0·01, 0·001, and 0·0001 centred at the sink phase, one can reach any phase outside the region of width 0·032 around the source phase, by transferring it four, five, and six times respectively. If we allow fifteen transfers, we may reach any phase outside the interval of width 10^{-8} around the source phase by choosing the initial phase in the region of width 10^{-4} centred at the sink phase.

The convergence towards the sink phase becomes slower as either β or $Q^{(1)}$ decreases. Each curve in Fig. 4.2 gives the lower bound of the region in which the rate of convergence $r(0{\cdot}02, 0{\cdot}005, \beta)$ is less than the value indicated at that curve. The dotted curve corresponds to the impurity frequency produced by one light atom with the mass $m^{(b)}$ imbedded in the regular chain. It is seen that for relatively large values of $Q^{(1)}$ the convergence is rapid in a major part of the region outside the band of the regular lattice. Fig. 4.3 shows how the rate of convergence changes as a

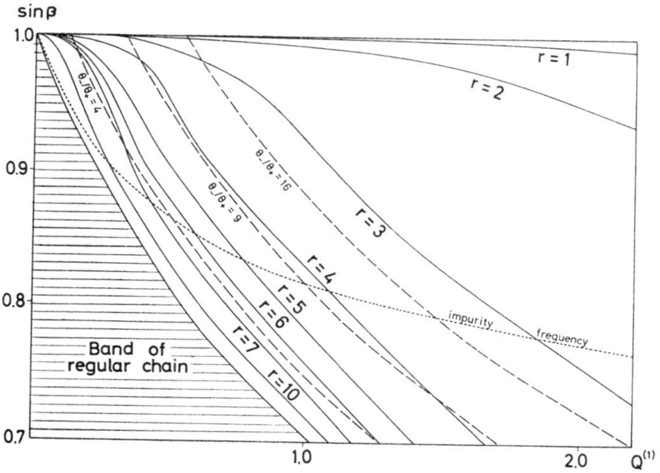

FIG. 4.2. The rate of convergence $r(0{\cdot}02, 0{\cdot}005, \beta)$ and the value of θ_-/θ_+ as functions of β and $Q^{(1)}$. The dotted curve gives the impurity frequency associated with one light atom with mass $m^{(b)}$ in the regular chain.

function of Δ_+ and Δ_- for the case $Q^{(1)} = 1$ and $\beta = 0{\cdot}955$. The broken line shows the variation of the rate under the condition $\Delta_+ = \Delta_-$.

It may be expected from (3.3.4) and (3.3.5) that the value of θ_-/θ_+ is a good measure for the rate of convergence, provided the angle between the limit vectors is not too small. The broken curves in Fig. 4.2 are isotimic curves of this quantity. It is seen that the above expectation is actually true. The large deviation of the broken from the full curve in the region $\beta \sim \pi/2$ is due to the proximity of the sink to the source vector, which in turn is a consequence of the singularity of the present representation at $\beta = \pi/2$. The extraordinary rapid convergence in the region $\beta \sim \pi/2$ is therefore merely an apparent one. If we used the representation without the singularity, the limit vectors would be well apart from each other

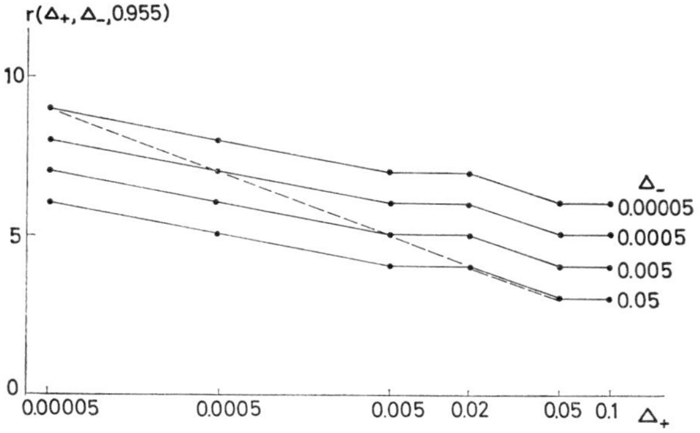

FIG. 4.3. The rate of convergence $r(\Delta_+, \Delta_-, 0{\cdot}955)$ as a function of Δ_+ and Δ_-.

as long as the eigenvalues were distinct, so that the rate of convergence would not become particularly rapid at $\beta = \pi/2$. Thus the true rate of convergence should rather be measured by θ_-/θ_+, which is independent of the representation. The result of computation by Hori and Fukushima is significant only in the region in which the full and broken curves run fairly parallel with one another.

Next turn to the consideration of the variation of the length of state vector during the transferring process. Here we assume that the transfer matrix is unimodular, i.e. det $\mathbf{Q} = 1$. This condition is satisfied in most cases. Then we have $|\theta_+| < 1$ and $|\theta_-| > 1$. If initially the state vector lies in a sufficiently narrow neighbourhood of the source vector, the coefficient c_+ in (3.3.4) will be much larger than c_- as long as the angle between the limit vectors is not too small. Therefore the inequality $|c_+\theta_+| \gg |c_-\theta_-|$ is valid, so that the lengths of $\mathbf{X}$ and the next vector $\mathbf{X}'$ are essentially equal to those of $c_+\mathbf{X}_+$ and $c_+\theta_+\mathbf{X}_+$, respectively. Since $|\theta_+| < 1$, the vector must be shortened. On the other hand, a vector in a neighbourhood of the sink vector is lengthened by a transfer. Hence in the process of successive transfers, any vector becomes shorter and shorter as long as it is in the neighbourhood of the source vector, while it becomes longer and longer during the approach to the sink vector. Since $|\theta_+|\,|\theta_-| = 1$, the rates of shortening and lengthening will be measured again by the ratio θ_-/θ_+. In consideration of the above results as regards the rate of convergence, it is expected that the change of vector length is likewise very rapid in general.

Figure 4.4 shows the variations of the vector length calculated by Hori and Fukushima for the same system as above. Figures indicated at each curve are the values of initial phases. It is seen

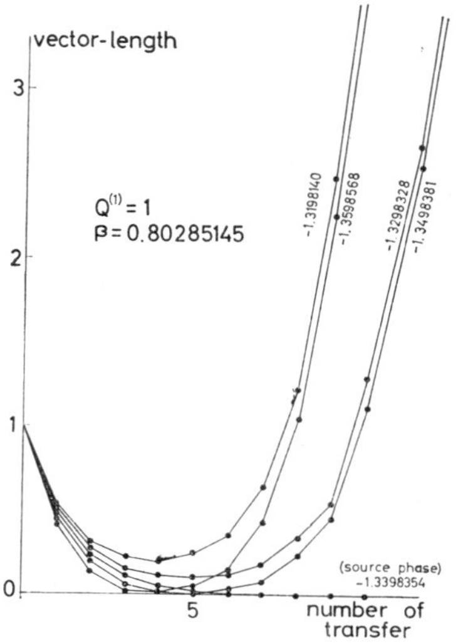

FIG. 4.4. The rate of variation of the vector length for $Q^{(1)} = 1$ and $\beta = 0{\cdot}80285145$. The values of the initial phase are indicated at each curve.

that the rate of increase or decrease of the length is actually remarkably large.

In the above the rapidity of the convergence of phase and the change of vector length was demonstrated upon the basis of a specific example. The point is, however, that both the rate of convergence of the phase and the rate of change of vector length are determined by the ratio θ_-/θ_+, and this does not depend on any particular model. Thus such a rapidity is well expected in general, in so far as the ratio θ_-/θ_+ is sufficiently large.

4.2. General Description of Impurity Modes

Suppose that in a regular system described by a transfer matrix $\mathbf{Q}^{(1)}$ there exists an impurity site or impurity region described by the matrix $\mathbf{Q}^{(0)}$. Let us start the transferring process from one end of the system where a boundary condition is imposed. It was shown in § 1.5 that to impose the boundary condition is equivalent to giving a specific value to the initial phase. Now suppose that $\mathbf{Q}^{(1)}$ is hyperbolic for the λ-value under consideration. If the boundary value of the phase corresponding to the boundary condition lies outside the narrow interval $(\delta_+^{(1)} - \Delta_+, \delta_+^{(1)} + \Delta_+)$, the phase comes into the interval $(\delta_-^{(1)} - \Delta_-, \delta_-^{(1)} + \Delta_-)$ after at most $r(\Delta_+, \Delta_-, \lambda)$ transfers. If the distance of the impurity site from the boundary is larger than $r(\Delta_+, \Delta_-, \lambda)$, the phase at the impurity site δ'_- is in the interval $(\delta_-^{(1)} - \Delta_-, \delta_-^{(1)} + \Delta_-)$. In case there is no impurity, the phase merely continues to approach $\delta_-^{(1)}$ by further transfers, so that it cannot satisfy the boundary condition at the terminal of the system unless the phase corresponding to the boundary condition at the terminal accidentally lies in the interval $(\delta_-^{(1)} - \Delta''_-, \delta_-^{(1)} + \Delta''_-)$, with $\Delta''_- \leqq \Delta_-$. Thus the value of λ now under consideration cannot give any eigenvalue. On the contrary, if the impurity exists, there is a possibility that δ'_- is transferred by $\mathbf{Q}^{(0)}$ to an appropriate value δ'_+ in the interval $(\delta_+^{(1)} - \Delta'_+, \delta_+^{(1)} + \Delta'_+)$, so that the phase reaches any boundary value outside the interval $(\delta_-^{(1)} - \Delta'_-, \delta_-^{(1)} + \Delta'_-)$, provided the distance between the impurity and the terminal is larger than $r(\Delta'_-, \Delta'_+, \lambda)$. Then our value of λ gives an eigenvalue λ_{imp}. Since $\mathbf{Q}^{(1)}$ is hyperbolic, λ_{imp} must lie outside the band of the regular lattice. The impurity frequency or the electronic impurity level mentioned in §§ 2.4 and 2.5 corresponds to such an eigenvalue.

If the extent of the system is infinite, the phase reaches just $\delta_-^{(1)}$ at the impurity irrespective of the starting boundary value. If $\delta_-^{(1)}$ is transferred by $\mathbf{Q}^{(0)}$ to $\delta_+^{(1)}$, the phase can reach any boundary value at the terminal, so that the λ- value under consideration gives an eigenvalue $\lambda_{\text{imp}}^{\infty}$. Since the difference between $\delta'_+ - \delta'_-$ and $\delta_+ - \delta_-$ is at most $2(\Delta_- + \Delta'_+)$ which is very small, one can make $\mathbf{Q}^{(0)}$ transform δ'_- to δ'_+ by adjusting the λ-value only very slightly. Thus λ_{imp} can differ from $\lambda_{\text{imp}}^{\infty}$ only by an extremely small amount, as far as the distances between the impurity and both ends of the system are sufficiently large, i.e. larger than max $\{r(\Delta_+, \Delta_-, \lambda), r(\Delta'_-, \Delta'_+, \lambda)\}$. Since, as was seen in the preceding section, the values of $r(\Delta_+, \Delta_-, \lambda)$ and $r(\Delta'_-, \Delta'_+, \lambda)$ are in general small integers, this condition may usually be considered to be fulfilled.

We must examine the above condition for the boundary values that the phases at the starting end and at the terminal should lie outside the intervals $(\delta_+^{(1)} - \Delta_+, \delta_+^{(1)} + \Delta_+)$ and $(\delta_-^{(1)} - \Delta''_-, \delta_-^{(1)} + \Delta''_-)$ respectively. Now that the values of $\delta_\pm^{(1)}$ depend on λ, whereas the usual physical boundary conditions are independent of the frequency or energy, the above condition cannot be violated throughout a finite range of λ-value. There still remains the possibility that it is violated in a very small interval at a special value of λ. Such a situation can occur, however, only accidentally, so that we may regard it as exceptional and omit from the general considerations.

According to the result obtained in the preceding section, the state vector becomes longer and longer as it is approaching the impurity and shorter and shorter as it is leaving it. Exceptions can occur only in two end regions of the system. In other words, the eigenmode corresponding to the eigenvalue λ_{imp}, that is, the mode corresponding to an impurity frequency or an electronic impurity level, is strongly localized around the impurity. Let us call such a mode *impurity mode* or *impurity state*. Obviously the amplitude of an impurity mode must decay towards either direction like a geometric series. This corresponds to an exponential decay in a continuous medium.

From the above argument it can be seen that the value of λ_{imp} is almost independent of both the boundary condition and the length of the system, provided the impurity is sufficiently far from both ends. Considering this, we shall assume in the following arguments that the system is infinite towards both directions in order to avoid unnecessary complications pertaining to the end effects.

Let the value of λ increase continuously from $\lambda_{imp} - \Delta\lambda$ to $\lambda_{imp} + \Delta\lambda$, where $\Delta\lambda$ is sufficiently small so that there is no other eigenvalue in this interval. The manner of the spatial variation of the phase δ changes only by an infinitesimal amount except at and immediately behind the impurity, since the phase is locked except in this region. Consequently, the total amount of the change of phase must increase by just 2π *within* this region, during the increase of λ by the amount $2\Delta\lambda$ through λ_{imp}. There are several possibilities in the manner of such an increase. Before the impurity, the phase remains always at $\delta_-^{(1)}$. Assume, for example, that the phase actually increases by 2π during one transfer by $\mathbf{Q}^{(1)}$, while the phase change by $\mathbf{Q}^{(0)}$ is smaller than 2π. Figure 4.5 illustrates one of the possible behaviours of phase for this case. The circular arrows indicate the changes of phase during each transfer. The thick arrow corresponds to the transfer by the impurity matrix $\mathbf{Q}^{(0)}$. For

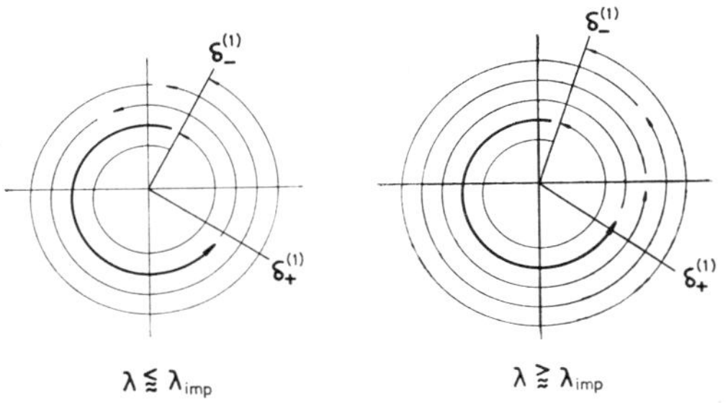

FIG. 4.5. A possible behaviour of the phase in the neighbourhood of λ_{imp}. Each circular arrow indicates the change of phase during one transfer. The thick arrow corresponds to the transfer by the impurity matrix $\mathbf{Q}^{(0)}$.

a λ-value slightly smaller than λ_{imp}, the phase changes at the impurity from $\delta_-^{(1)}$ to a value slightly smaller than $\delta_+^{(1)}$, while for a λ-value slightly larger than λ_{imp}, the phase becomes slightly larger than $\delta_+^{(1)}$. It is seen that, when λ remains smaller than λ_{imp}, the advance of the phase is always retarded by the impurity, but from the instant at which λ passes λ_{imp}, the phase abruptly begins to be accelerated, to increase the total change by just 2π. According to the terminology of Roberts and Makinson (1962), we call this phenomenon the *slip* of the phase.

We had better turn here to the examination of some concrete examples instead of investigating the various possibilities for the slip in an abstract manner.

4.3. Vibrational Impurity Modes

At first we consider the vibrational impurity mode produced by an isolated isotopic impurity. Let the masses of the impurity and the host atom be $m^{(0)}$ and $m^{(1)}$ respectively. If we use the impurity atom as the basic one, i.e. if we put $m^{(b)} = m^{(0)}$, the transfer matrices at the impurity and the host atom are given by

$$\mathbf{Q}^{(0)} = \begin{pmatrix} \exp(2i\beta) & 0 \\ 0 & \exp(-2i\beta) \end{pmatrix}, \tag{4.3.1}$$

and (3.2.8) with $i = 1$ respectively, where now $Q^{(1)} \equiv (m^{(1)}/m^{(0)}) - 1$.

For real values of β, $\mathbf{Q}^{(0)}$ is evidently elliptic. For an impurity mode to appear, $\mathbf{Q}^{(1)}$ must be hyperbolic, that is, the frequency must be in the spectral gap of the regular chain. The eigenvalue equation of $\mathbf{Q}^{(1)}$ and its eigenvalues are given by

$$\theta^2 - 2A^{(1)}(\beta)\,\theta + 1 = 0 \tag{4.3.2}$$

and

$$\theta_{\pm} = A^{(1)}(\beta) \pm \{A^{(1)2}(\beta) - 1\}^{1/2} \tag{4.3.3}$$

respectively, where

$$A^{(1)}(\beta) \equiv \cos 2\beta - \tan\gamma^{(1)} \sin 2\beta . \tag{4.3.4}$$

The condition for $\mathbf{Q}^{(1)}$ to be hyperbolic is given by

$$A^{(1)}(\beta) < -1 . \tag{4.3.5}$$

The fact that $A^{(1)}(\beta)$ is always negative for real $\theta_{\pm}^{(1)}$ has been taken into account in order to give the double sign in (4.3.3) the right order, so that the larger modulus may be given to $\theta_{-}^{(1)}$. From (4.3.4) it is seen that (4.3.5) is equivalent to

$$\sin^2\beta > m^{(0)}/m^{(1)} . \tag{4.3.6}$$

The value of β given by $\sin^2\beta = m^{(0)}/m^{(1)}$ corresponds to the maximum frequency (the top of the band) of the regular lattice for given $m^{(0)}$. Equation (4.3.6) means that for any given value of

β, the impurity atom must necessarily be lighter than the host atom if the impurity mode is to appear.

At the top of the band, i.e. when $A^{(1)}(\beta) = -1$, we have $\theta_{\pm}^{(1)} = -1$. As β increases to $\pi/2$, $\theta_{-}^{(1)}$ decreases down to $-1 - 2Q^{(1)} - \{(1 + 2Q^{(1)})^2 - 1\}^{1/2}$, while $\theta_{+}^{(1)}$ increases up to $-1 - 2Q^{(1)} + \{(1 + 2Q^{(1)})^2 - 1\}^{1/2}$.

We obtain for the limit points $z_{\pm}^{(1)}$ the expression

$$z_{\pm}^{(1)} = \{(1 - i \tan \gamma^{(1)}) \exp(-2i\beta) - \theta_{\pm}^{(1)}\}/i \tan \gamma^{(1)} \exp(2i\beta), \tag{4.3.7}$$

and for the limit phases $\delta_{\pm}^{(1)}$

$$\cos(\delta_{\pm}^{(1)}/2) = \left\{\frac{\sin 2\beta}{-2\theta_{\pm}^{(1)} \tan \gamma^{(1)}}\right\}^{1/2}. \tag{4.3.8}$$

Differentiating the square of the right-hand side of (4.3.8), we have

$$\frac{d \cos^2(\delta_{\pm}^{(1)}/2)}{d\beta} = \frac{-\sin 2\beta}{2 \tan \gamma^{(1)}} \cdot \frac{d\theta_{\mp}^{(1)}}{d\beta} + \frac{(1 - \cos 2\beta)\theta_{\mp}^{(1)}}{\tan \gamma^{(1)}}, \tag{4.3.9}$$

which is always negative for the lower sign, and for the upper sign is at first positive and then becomes negative. For $\theta_{\pm}^{(1)} = -1$ we get

$$\cos(\delta_{\pm}^{(1)}/2) = \{1/(Q^{(1)} + 1)\}^{1/2}, \tag{4.3.10}$$

and for $\beta \to \pi/2$,

$$\cos(\delta_{\pm}^{(1)}/2) \to 0. \tag{4.3.11}$$

Thus it is seen that the source and the sink phase $\delta_{\pm}^{(1)}$ both start from the value $2\cos^{-1}\{1/(Q^{(1)} + 1)\}^{1/2}$, but the former decreases down to $-\pi$, while the latter increases up to $+\pi$ (Fig. 4.6).

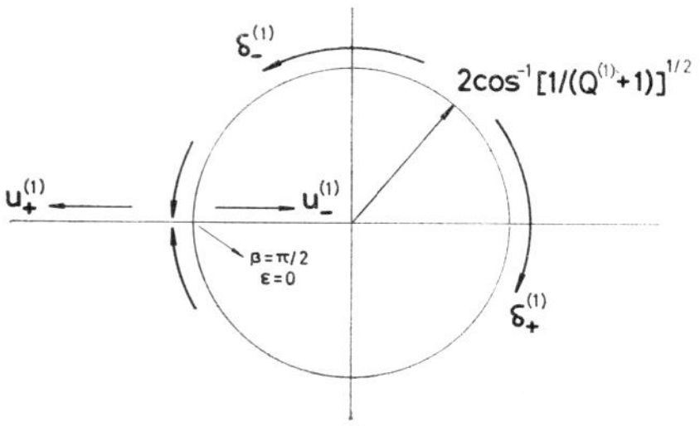

FIG. 4.6. The variation of the limit phases $\delta_{\pm}^{(1)}$ and the limit state ratios $u_{\pm}^{(1)}$ with the increase of β and ε, respectively.

For frequencies larger than $\omega = 2(K/m^{(0)})^{1/2}$, which corresponds to $\beta = \pi/2$, we have to put $\beta = (\pi/2) + i\varepsilon$. Then the transfer matrix and $A^{(1)}(\beta)$ become a real matrix

$$\mathbf{T}^{(1)} = \begin{pmatrix} -(1 - Q^{(1)} \coth \varepsilon) \exp(-2\varepsilon) & Q^{(1)} \coth \varepsilon \exp(2\varepsilon) \\ -Q^{(1)} \coth \varepsilon \exp(-2\varepsilon) & -(1 + Q^{(1)} \coth \varepsilon) \exp(2\varepsilon) \end{pmatrix}, \tag{4.3.12}$$

and

$$A^{(1)}(\varepsilon) = -\cosh 2\varepsilon - Q^{(1)} \coth \varepsilon \sinh 2\varepsilon \tag{4.3.13}$$

respectively. Since $A^{(1)}(\varepsilon)$ is always less than -1, $\mathbf{T}^{(1)}$ is always hyperbolic, as it should be.

The eigenvalues and the limit points of $\mathbf{T}^{(1)}$ are given by

$$\theta_{\pm}^{(1)} = A^{(1)}(\varepsilon) \pm \{A^{(1)2} - 1\}^{1/2} \tag{4.3.14}$$

and

$$u_{\pm}^{(1)} = -\{\theta_{\pm}^{(1)} + (1 + Q^{(1)} \coth \varepsilon) \exp(2\varepsilon)\}/Q^{(1)} \coth \varepsilon \exp(-2\varepsilon) \tag{4.3.15}$$

respectively. The limiting values of $u_{\pm}^{(1)}$ for $\varepsilon \to \infty$ are easily seen to be

$$\lim_{\varepsilon \to \infty} u_{+}^{(1)} = -\infty \quad \text{and} \quad \lim_{\varepsilon \to \infty} u_{-}^{(1)} = 0. \tag{4.3.16}$$

Combining the behaviours of $\delta_{\pm}^{(1)}$ and $u_{\pm}^{(1)}$, we have the result shown in Fig. 4.6. The coincidence of the sink with the source phase at $\beta = \pi/2$ ($\varepsilon = 0$) is a consequence of the singularity of the transfer matrix now being used.

The condition for $z_{-}^{(1)}$ to be transferred to $z_{+}^{(1)}$ by $\mathbf{Q}^{(0)}$ is

$$(1 - i \tan \gamma^{(1)}) - \theta_{-}^{(1)} \exp(2i\beta) = (1 - i \tan \gamma^{(1)}) \exp(-4i\beta) - \theta_{+}^{(1)} \exp(-2i\beta), \tag{4.3.17}$$

which can be reduced to

$$2 \tan \gamma^{(1)} = -\tan 2\beta. \tag{4.3.18}$$

This is the equation for the impurity frequency, which coincides with that derived by Montroll and Potts (1955) and Hori and Asahi (1957).

As was seen above, at the frequency corresponding to the band top of the regular lattice, we have $\theta_{\pm}^{(1)} = -1$, so that the state vector reverses its direction during each transfer by $\mathbf{Q}^{(1)}$, once it has reached the sink vector. This means that before reaching the impurity the phase increases by the amount $2n\pi$ (n: integer) at this frequency. Since in the present case the equation governing the system is a difference equation of the type (3.1.1), the increase of the phase during one transfer is at most 2π. Hence the integer n must be unity. At the impurity site, however, the phase increases only by the amount $4\beta < 2\pi$. This gives rise to a retardation of phase, owing to which the frequency, which would be the maximum frequency if the lattice were regular, fails to become an eigenfrequency. If the frequency is increased, the phase is locked owing to the hyperbolicity of $\mathbf{Q}^{(1)}$, so that the phase increase during one transfer must remain to be 2π. The amount of phase increase at the impurity site becomes larger, however. When the frequency reaches the impurity frequency given by (4.3.18), the phase increase 4β at the impurity becomes so large that the next phase reaches the source phase $\delta_+^{(1)}$. Then a slip occurs, the phase suddenly begins to be accelerated, and at length one has an eigenfrequency. In other words, the maximum eigenfrequency has been "pushed up" from the band top to the position of the impurity frequency by the impurity atom. From the above argument and the variation of $\delta_\pm^{(1)}$ shown in Fig. 4.6, it is seen that in the present case the type of the slip is just as shown in Fig. 4.5.

The manner of appearance of the impurity frequencies in the case of two isotopic impurities may be explained in a similar way. A possibility is shown in Fig. 4.7 for the case in which the interval between impurities is four, which is assumed to be sufficiently large, that is, larger than the rate of convergence $r(\Delta_+, \Delta_-, \lambda)$ with appropriately small Δ_+ and Δ_-. The two eigenvalues are denoted by $\lambda_{\text{imp}}^{(1)}$ and $\lambda_{\text{imp}}^{(2)}$ ($\lambda_{\text{imp}}^{(1)} < \lambda_{\text{imp}}^{(2)}$). It is seen that these eigenvalues are very close to the eigenvalue λ_{imp} for the case of one impurity. The difference between two eigenvalues comes from a very subtle difference in the behaviour of the phase. As long as the impurities are sufficiently apart from each other, two slips occur successively at two close-lying λ-values.

The situation remains essentially the same, if there are many impurities provided that they are sufficiently apart from one another. Within a narrow interval of λ, many slips occur successively at different impurity atoms, and correspondingly there appear a large number of eigenfrequencies around the impurity frequency of an isolated impurity. From the above considerations it may be concluded that if the distance between impurities is larger than $r(\Delta_+, \Delta_-, \lambda)$, in other words, if the fraction of impurity atoms is less than $1/r$, these eigenfrequencies will form a narrow impurity band whose width is of the order max (Δ_+, Δ_-). Note that r is ordinarily a small integer. This provides us with the explanation of the peak structure in the spectrum of disordered chain, as will be shown in detail in § 5.6.

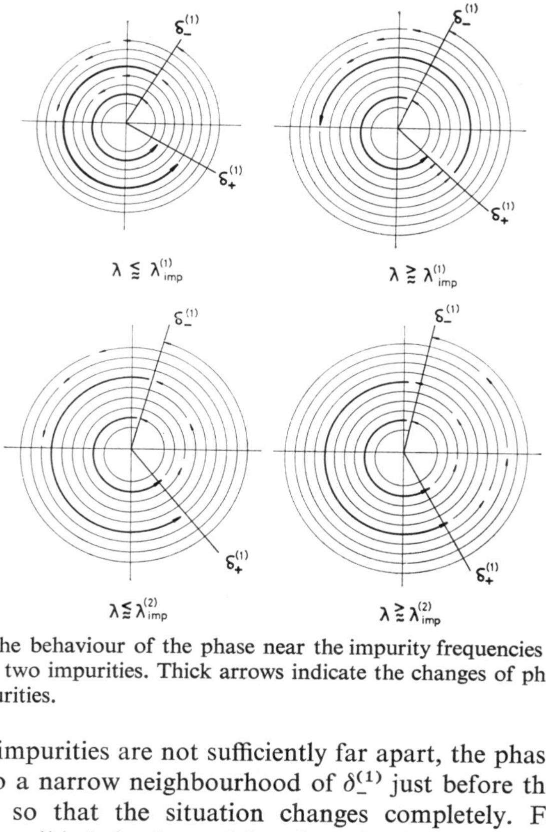

FIG. 4.7. The behaviour of the phase near the impurity frequencies for the case of two impurities. Thick arrows indicate the changes of phase at the impurities.

If two impurities are not sufficiently far apart, the phase fails to come into a narrow neighbourhood of $\delta_-^{(1)}$ just before the second impurity, so that the situation changes completely. Figure 4.8 shows a possible behaviour of the phase for the case of two neighbouring impurities. Here the two eigenvalues still exist outside the band, but one of them is rather larger while the other is rather smaller than those in the previous case. The lower eigenvalue may therefore dip into the band of regular lattice. In such a case $\mathbf{Q}^{(1)}$ is elliptic at this frequency, so that the present manner of description fails, and only the higher eigenvalue remains to correspond to an impurity mode.

The total phase change due to two neighbouring impurities is equal to 2π at the band top when

$$Q^{(1)} = 1. \tag{4.3.19}$$

This is just the condition for the value $\beta = \pi/4$ to correspond to the band top, and means that the lower impurity frequency does not exist for $m^{(1)}/m^{(0)} < 2$. This result coincides with that obtained by Takeno, Kashiwamura, and Teramoto (1963).

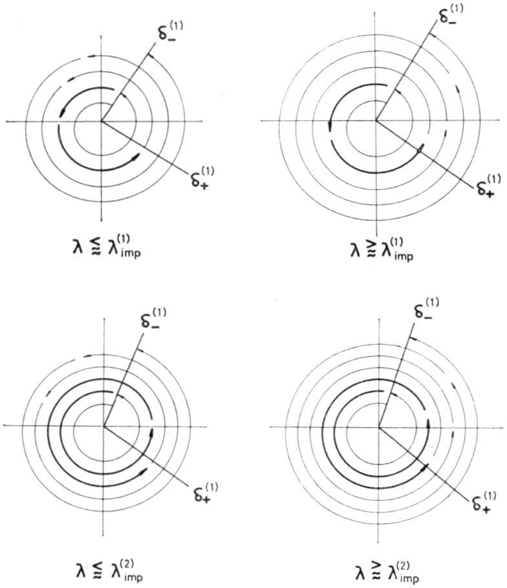

FIG. 4.8. The behaviour of the phase near the impurity frequencies for the case of two neighbouring impurities.

4.4. Behaviour of Limit Phases in the Regular Kronig–Penney Model

The transfer matrix for a regular array of δ-potentials of strength $-V$, with spacing l, is given by (3.4.2) or (1.3.20) without the index n. Here it is convenient to use the latter matrix $\mathbf{T}$ to calculate the positions of limit phases, and then turn to a complex representation by an appropriate Cayley-type transformation. The matrix

(3.4.2) is inconvenient, notwithstanding its simple and symmetric form, due to its singularity at $k = 0$. This causes considerable complications in the behaviour of the phase in the neighbourhood of $k = 0$ and in the negative-energy region.

The eigenvalue equation and eigenvalues of $\mathbf{T}$ are

$$\theta^2 - 2A(k)\,\theta + 1 = 0 \tag{4.4.1}$$

and

$$\theta_+ = A(k) \pm \{A^2(k) - 1\}^{1/2}, \tag{4.4.2a}$$

$$\theta_- = A(k) \mp \{A^2(k) - 1\}^{1/2} \tag{4.4.2b}$$

respectively. The condition for $\theta_\pm$ to be real is that $|A(k)| > 1$. In contrast to the vibrational case, in which the corresponding quantity $A(\beta)$ cannot be larger than unity, both possibilities $A(k) > 1$ and $A(k) < -1$ may now be realized. The odd gaps correspond to the case $A(k) < -1$ and even gaps to $A(k) > 1$. The eigenvalues $\theta_\pm$ are negative or positive according as $A(k) < -1$ or >1, so that we should adopt upper or lower sign in (4.4.2), according as $A(k) < -1$ or $A(k) > 1$. Further, the sign of $\theta_\pm$ is opposite to that of $\sin kl$ for $V > 0$, while for $V < 0$ these signs agree with each other.

The limit points $u_\pm$ are given by

$$u_\pm = \frac{\cos kl - (V/k)\sin kl - \theta_\pm}{k \sin kl + V \cos kl} = \frac{-(V/2k)\sin kl_\pm\{A^2(k) - 1\}^{1/2}}{B(k)}, \tag{4.4.3}$$

where $B(k) \equiv k \sin kl + V \cos kl$. The correspondence between the double signs in the second and the third expressions must be properly chosen according to the sign of $A(k)$.

At $kl = n\pi$, where n is a positive integer, both u_+ and u_- vanish:

$$u_+ = u_- = 0, \tag{4.4.4}$$

while at $A(k) = \pm 1$, i.e. at the bottoms of the gaps for $V > 0$ and at the tops of the gaps for $V < 0$, they become

$$u_\pm = -(V/2k)\sin kl/B(k). \tag{4.4.5}$$

The zeros of the function $B(k)$ lie in the spectral gaps. For if $B(k) = 0$, $A^2(k)$ becomes

$$A^2(k) = 1 + (V^2/4k^2)\sin^2 kl, \tag{4.4.6}$$

which shows that $A^2(k) > 1$. Conversely, in every gap there is one zero of $B(k)$. This means that the values of $B(k)$ at the boundaries of each gap have opposite signs. Since at $kl = n\pi$ $B(k)$ is equal to $-V$ or V, according as n is odd or even, these signs are as shown in Table 4.1. Then from (4.4.5), the signs of $u_\pm$ at the boundaries can be obtained as given in the last line of the table.

TABLE 4.1 *Signs of $B(k)$ and $u_\pm$ at the boundaries of spectral gaps*

	$V > 0$				$V < 0$			
	Odd		Even		Odd		Even	
	Bottom	Top	Bottom	Top	Bottom	Top	Bottom	Top
Sign $B(k)$	+	−	−	+	+	−	−	+
Sign $u_\pm$	−	0	−	0	0	+	0	+

At $B(k) = 0$, the denominator of (4.4.3) vanishes, so that here either u_+ or u_- must pass the infinity. From (4.4.6), it is seen that the eigenvalues $\theta_\pm$ have at this point the values

$$\theta_+ = A(k) + (V/2k)\sin kl = \cos kl,$$
$$\theta_- = A(k) - (V/2k)\sin kl = \cos kl - (V/k)\sin kl. \tag{4.4.7}$$

Consequently the numerator of (4.4.3) vanishes for u_-, while equal to $-(V/k)\sin kl$ for u_+. Thus the state ratio which becomes infinity at $B(k) = 0$ is u_+ and not u_-.

By applying the transformation $z = (u - i)/(u + i)$, the real axis is mapped on to the unit circle of the complex plane as shown in Fig. 4.9a. Hence the variation of the limit phases with the increase of k becomes as shown in Fig. 4.9b, c.

The zeroth gap for $V < 0$ extends to the negative-energy region. The first gap for $V > 0$ also continues to the negative-energy region when $Vl > 4$. For negative energies we must put $k = i\varkappa$. Then all the trigonometric functions in the above formulae are replaced by the corresponding hyperbolic functions except that the function $B(k)$ must be replaced by

$$B(\varkappa) \equiv -\varkappa \sinh \varkappa l + V \cosh \varkappa l. \tag{4.4.8}$$

Consider at first the zeroth gap for $V < 0$. At $\varkappa = 0$, the state ratios have the value

$$u_\pm = \frac{1}{V}\left[-\frac{Vl}{2} \pm \left\{\left(\frac{Vl}{2}\right)^2 - Vl\right\}^{1/2}\right]. \tag{4.4.9}$$

This shows that u_+ is negative while u_- is positive, so that u_+ passes the infinity in the positive-energy region. This is as it should be, since from (4.4.8) $B(\varkappa)$ does not change its sign in the negative-energy region. It is readily seen that for $\varkappa \to \infty$, both u_+ and u_- become zero. Thus the variation of the limit phases in this gap is the same as in the higher gaps.

For the first gap for $V > 0$, we have at $\varkappa = 0$

$$u_\pm = \frac{1}{V}\left[-\frac{Vl}{2} \mp \left\{\left(\frac{Vl}{2}\right)^2 - Vl\right\}^{1/2}\right], \qquad (4.4.10)$$

which shows that here both u_+ and u_- are negative. Since $A(\varkappa)$ cannot be negative at the root of $B(\varkappa) = 0$ in the negative-energy region, so that $B(\varkappa)$ cannot vanish within the negative-energy part of the first gap, this means that u_+ must pass the infinity in the positive-energy part, and both u_+ and u_- must be negative through-

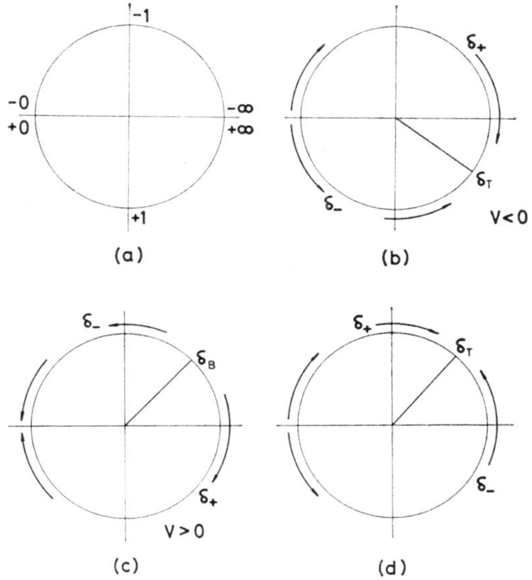

FIG. 4.9. (a) The correspondence between the real axis and the unit circle through the mapping $z = (u - i)/(u + i)$. (b) (c) The variation of limit phases in each gap with the increase of k for $V < 0$ and $V > 0$ respectively. (d) The variation of the limit phases in the zeroth gap for $V > 0$. δ_B and δ_T are the phases at the bottom and the top of the gap respectively.

out the negative-energy portion of the first gap. The variation of the limit phases is thus again the same as in the higher gaps.

Finally, consider the zeroth gap for $V > 0$, which lies wholly in the negative-energy region. By a similar argument to above, we can conclude that $u_\pm$ are negative at the top of the gap, both vanish for $\varkappa \to \infty$, and in the meantime u_- passes the infinity. Thus the variation of the limit phases becomes as shown in Fig. 4.9d.

It is to be noted that in every case the sink phase δ_- increases, while the source phase δ_+ decreases, as k increases or $\varkappa$ decreases.

4.5. Impurity Levels in the Kronig–Penney Model

Suppose that there exists an impurity potential with strength $-V^{(0)}$ in the regular array of δ-potentials considered in the preceding section. The transfer matrix $\mathbf{T}^{(0)}$ associated with the cell containing the impurity potential is given by (1.3.20) with $V_n = V^{(0)}$ and $l_n = l$. The condition for the impurity level is, as in the vibrational case, that the sink vector $\mathbf{X}_-$ of $\mathbf{T}$ is transferred to the source vector $\mathbf{X}_+$ of $\mathbf{T}$ by the matrix $\mathbf{T}^{(0)}$:

$$\mathbf{T}^{(0)}\mathbf{X}_- = \mathbf{X}_+. \tag{4.5.1}$$

After some calculations this is reduced to

$$A^{(0)}(k) \equiv \cos kl - (V^{(0)}/2k) \sin kl = \theta_+. \tag{4.5.2}$$

Substituting (4.4.2), and considering that the signs of $A(k)$ and $\sin kl$ are always opposite for $V > 0$, while they agree for $V < 0$, we get

$$V^{(0)} = V \pm \frac{2k}{|\sin kl|} \left\{ \left(\cos kl - \frac{V}{2k} \sin kl \right)^2 - 1 \right\}^{1/2}, \tag{4.5.3}$$

where we should take the plus sign for $V < 0$ and minus sign for $V > 0$. This equation shows that, in case the host lattice is composed of potential "peaks" ($V < 0$), there exists an impurity state when $V^{(0)} > V$, that is, when the impurity potential is a lower peak or is a "trough". On the other hand, in case the host potentials are troughs ($V > 0$), there appears an impurity level when $V^{(0)} < V$, that is, when the impurity potential is a shallower trough or is a peak.

For negative energies, we have only to put $k = i\varkappa$ and replace the trigonometric functions by the corresponding hyperbolic functions. For $V < 0$, $\mathbf{T}$ is always hyperbolic, and the plus sign should be adopted in (4.5.3). No negative $V^{(0)}$ can satisfy (4.5.2) in the negative-energy region, since the left-hand side is always larger while the right-hand side always smaller than unity. A positive $V^{(0)}$ may, however, satisfy (4.5.3) and give rise to an impurity level, if it is sufficiently large so that θ_+ becomes equal to $A^{(0)}(k)$ in the negative-energy region. For $V > 0$, we should take the plus sign for the zeroth gap, and the minus sign for the first gap. In the first gap, there appears an impurity level if $V^{(0)} < V$, while in the zeroth gap an impurity level appears if $V^{(0)} > V$.

To sum up, in the case of $V < 0$, each gap contains an impurity level when $V^{(0)} > V$ ($V^{(0)}$ may be positive). On the contrary, in the case of $V > 0$, an impurity level appears in each gap when $V^{(0)} < V$ ($V^{(0)}$ may be negative), except the zeroth gap in which it appears when $V^{(0)} > V$. In Fig. 4.10, the conditions for the appearance of impurity states are schematically depicted, and the regions which contain an impurity state are indicated by thick lines.

The appearance of electronic impurity levels can also be explained by the slip of phase at the impurity potential in a way which is completely similar to that described in § 4.3. For this purpose, it is to be noted that the amount of change of the phase by $\mathbf{T}$ and $\mathbf{T}^{(0)}$ should be roughly equal to the phase changes per unit cell in the regular lattice and in the lattice composed of impurity potentials only, respectively. Precisely, it is approximately equal to zero, 4π, 4π, 8π, 8π, ... at the boundaries of the even gaps, while equal to 2π, 2π, 6π, 6π, ..., at those of the odd gaps, in the order of increasing energy, remaining constant throughout each gap.

For the zeroth gap, the amount of phase change by $\mathbf{T}$ is zero, while that by $\mathbf{T}^{(0)}$ is less than 2π, so that the variation of phase in the neighbourhood of the impurity level must be such as shown in Fig. 4.11. For the energy slightly smaller than the impurity level ($\lambda \lessapprox \lambda_{\mathrm{imp}}$), the phase is fixed at the sink phase δ_- of $\mathbf{T}$ until the impurity potential is reached, where it is advanced to the value slightly smaller than the source phase δ_+, so that henceforth it is pulled back by δ_-. Thus after all the total increase of phase remains zero, i.e. the same as in the case of absence of the impurity. For the energy slightly larger than the impurity level ($\lambda \gtrapprox \lambda_{\mathrm{imp}}$), however, the phase advances at the impurity a bit beyond δ_+. (The phase change by $\mathbf{T}^{(0)}$ increases while the angle between δ_- and δ_+ diminishes with λ.) Hence the phase advances further, until it

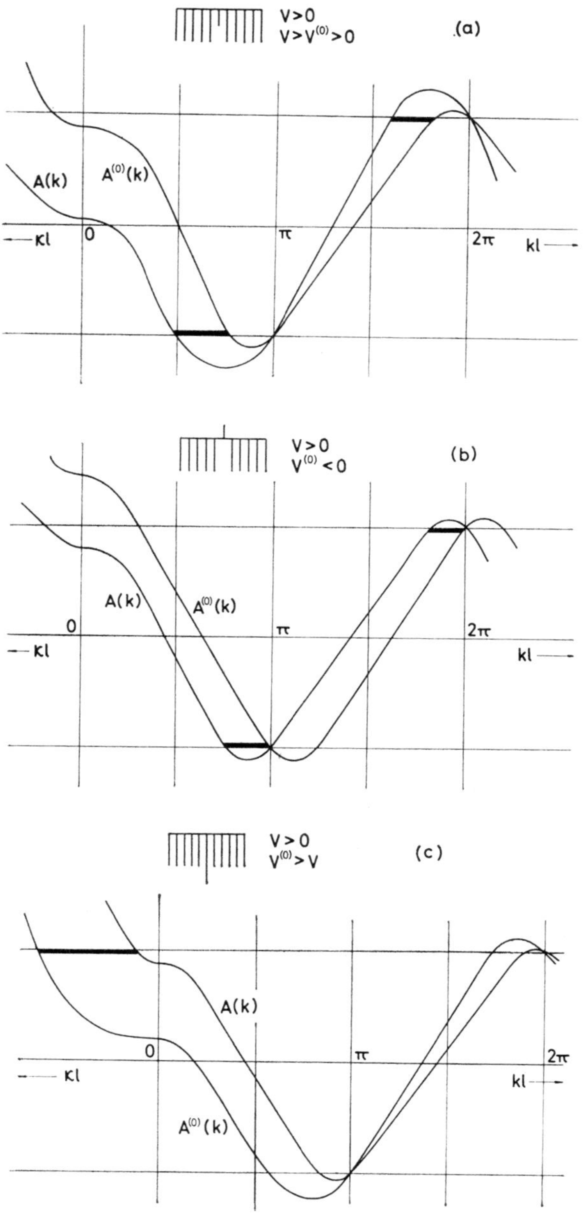
V>0
V>V^(0)>0
(a)
A(k)
A^(0)(k)
κl
0
π
2π
kl
V>0
V^(0)<0
(b)
A(k)
A^(0)(k)
0
π
2π
κl
kl
V>0
V^(0)>V
(c)
A(k)
0
π
2π
κl
kl
A^(0)(k)

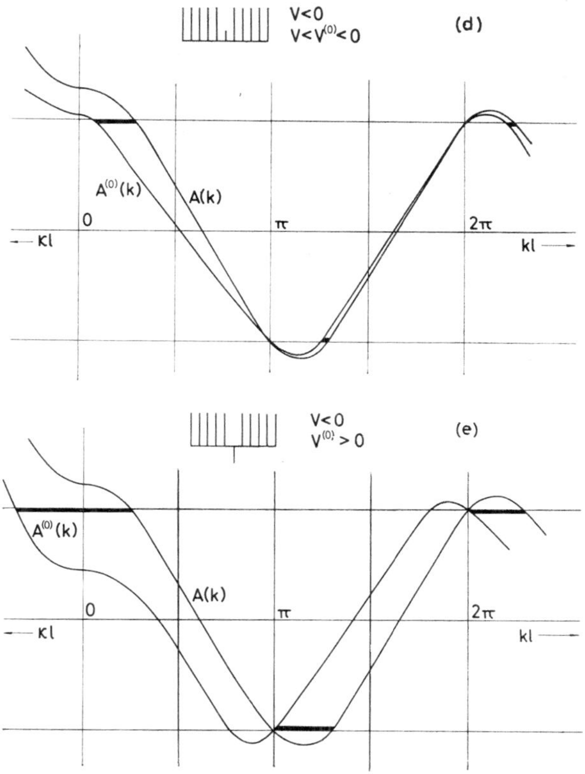

FIG. 4.10. The functions $A(k)$ and $A^{(0)}(k)$ are plotted for various conditions on the values of V and $V^{(0)}$. Each condition is schematically depicted above the corresponding graph by an array of vertical lines indicating the strength of potentials. For each condition, the energy regions indicated by thick lines contain an impurity level.

finally reaches δ_- from the opposite side, so that the total phase change abruptly increases by 2π. In other words, a slip occurs at the impurity, and the total amount of the phase change becomes 2π, i.e. the same as in the lowest state (the state at the bottom of the first band) of the regular lattice. The impurity has, so to speak, "pulled down" the lowest state from the bottom of the first band to make it an impurity state, by accelerating the phase advance and making the total phase change 2π already at an energy lower than the bottom of the first band. This is in harmony with the fact that the value of λ_{imp} approaches the bottom of the first band as $V^{(0)}$ approaches V. In other words, the impurity state sohuld be regarded to have "descended" from the bottom of the first band.

Next consider the first gap. For $V > 0$, the phase change by $\mathbf{T}$ is equal to 2π, while that by $\mathbf{T}^{(0)}$ is less than 2π. For $\lambda \lessapprox \lambda_{\text{imp}}$, the

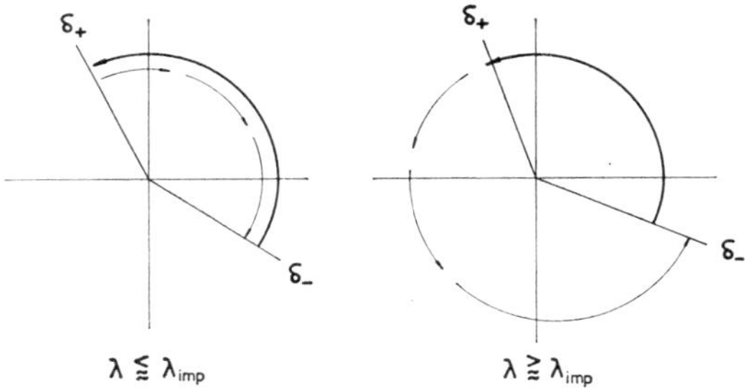

FIG. 4.11. The behaviour of the phase near the impurity level in the zeroth zap.

phase changes by just 2π during each transfer until the impurity is reached, where it advances to a value slightly smaller than δ_+, so that henceforth it is pulled back to δ_-, changing by an amount less than 2π during a transfer. After reaching δ_-, it again changes by just 2π per one transfer. For $\lambda \gtrapprox \lambda_{\text{imp}}$, on the other hand, the phase advances at the impurity up to the value slightly larger than δ_+, so that thenceforth it changes by an amount larger than 2π during each transfer, finally reaching δ_- from the opposite side. Its advance suddenly begins, therefore, to be accelerated at $\lambda = \lambda_{\text{imp}}$, and the total change of the phase is recovered to the value which the highest state of the first band would have if there were no impurity.

The situation in the first gap for $V > 0$ is illustrated in Fig. 4.12, in which the phase variation for the case of regular lattice is also shown. In the present case the impurity "pushes up" the impurity state from the top of the first band, by decelerating the phase advance, so that its total change reaches that of the state of the regular lattice at the top of the first band only at a higher energy. This is in harmony with the fact that λ_{imp} approaches the top of the first band as $V^{(0)}$ approaches V, and that at the top of the first gap we have $\mathbf{T} = \mathbf{T}^{(0)}$ and consequently the amount of the phase change due to **T** is the same as that due to $\mathbf{T}^{(0)}$. This means that the total phase change is fixed irrespective of whether the impurity is present or not. Now the impurity level should be considered to have "ascended" from the first band.

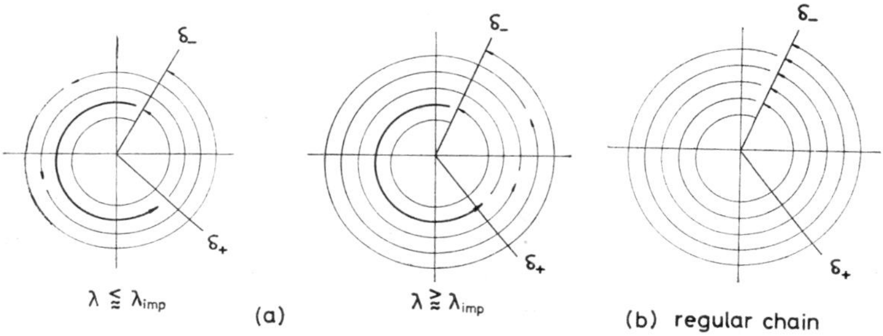

FIG. 4.12. The behaviour of the phase near the impurity level in the first gap for $V > 0$.

For $V < 0$, the phase change by **T** is equal to 2π while that by $\mathbf{T}^{(0)}$ is between 2π and 4π. It may readily be seen that the phase variation must be as shown in Fig. 4.13. The total phase change in

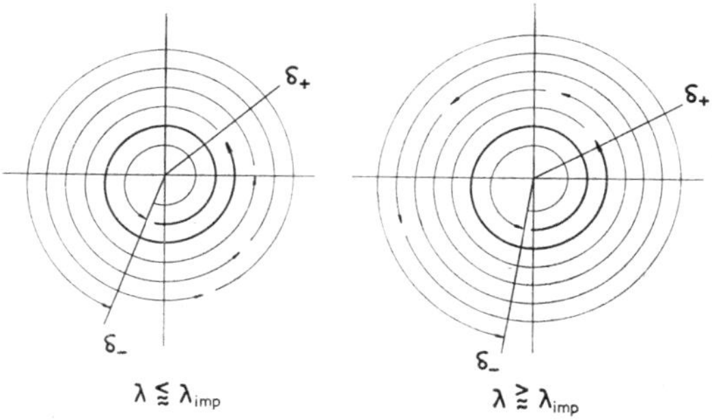

FIG. 4.13. The behaviour of the phase near the impurity level in the first gap for $V < 0$.

the impurity state is now equal to that at the bottom of the second band for the regular lattice. The impurity again "pulls down" the impurity level from the upper band by accelerating the phase advance, to make its total change reach the value at the bottom of the upper band already at a lower energy. That this ought to be the case is also seen from the fact that λ_{imp} approaches the bottom of the second band as $V^{(0)} \to V$, and that the total phase change is fixed at the top of the first band irrespective of whether there is an impurity or not.

By similar arguments, we obtain the conclusion that, if $V < 0$, the impurity level appears in every gap from the bottom of the band lying just above, while for $V > 0$, it appears from the top of the band lying just below, except the zeroth gap where it appears from the above band as in the case of $V < 0$. The phase variation in the vicinity of each impurity level is completely similar to those shown in Figs. 4.11, 4.12, and 4.13. The difference lies only in the amount of phase change during each transfer. For the second gap, it is twice as much as the change in the first gap; for the third gap, it is three times, and so on.

Finally, let us examine the case in which there is a cell with a length different from the regular spacing of potentials instead of a potential with a different strength. In this case it is convenient to use the representation (3.2.11). Then the transfer matrices at the regular and the impurity cells are given by (3.4.2) and (3.2.11) with $V_n = V$ and $l_n = l^{(0)}$ respectively, where $l^{(0)}$ is the length of the impurity cell. Denoting the latter matrix by $\mathbf{Q}^{(0)}$, we have the equation

$$\mathbf{Q}^{(0)}\mathbf{X}_- = \mathbf{Q}\begin{pmatrix} \exp\{ik(l^{(0)} - l)\} & 0 \\ 0 & \exp\{-ik(l^{(0)} - l)\} \end{pmatrix}\mathbf{X}_- = \mathbf{X}_+, \quad (4.5.4)$$

as the condition for the impurity state, where $\mathbf{X}_\pm$ are the limit vectors of the matrix $\mathbf{Q}$. This equation means that the operation of $\mathbf{Q}^{(0)}$ on $\mathbf{X}$ is decomposed into two steps: firstly, the sink vector $\mathbf{X}_-$ is transferred to the source vector $\mathbf{X}_+$ by a simple rotation by an angle $2k(l^{(0)} - l)$, and then it is transferred by $\mathbf{Q}$ to $\mathbf{X}_+$ itself. As the quantity $2k(l^{(0)} - l)$ can only be determined up to the difference of integral multiples of 2π, there are infinitely many values of $l^{(0)}$ which satisfy (4.5.4).

Equation (4.5.4) can readily be reduced to

$$\cos\{2k(l - l^{(0)})\} = \frac{4k^2}{V^2}\left\{2(1 - A^2(k)) + \frac{V^2}{4k^2}\right\}, \quad (4.5.5a)$$

$$\sin\{2k(l - l^{(0)})\} = \pm\frac{4k^2}{V^2}(A^2(k) - 1)^{1/2}\, 2C(k), \quad (4.5.5b)$$

where
$$C(k) \equiv \sin kl + (V/2k) \cos kl, \tag{4.5.6}$$
and the upper and lower signs correspond to the cases $A(k) < -1$ and $A(k) > 1$ respectively. In order to obtain the value of $l^{(0)}$ which gives rise to an impurity level for given l and k, we plot the right-hand sides of (4.5.5) as functions of l, and seek for each l the value of $l^{(0)}$ for which the values of these functions are consistently equal to $\cos 2kl$ and $\sin 2kl$ respectively (Fig. 4.14). It is seen that actually there exist infinitely many values of $l^{(0)}$, differing by $n\pi/k$ (n: integer) from one another, which satisfy this condition.

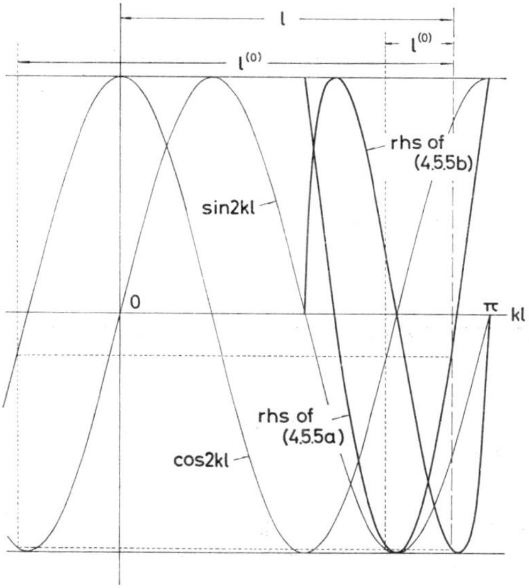

FIG. 4.14. The method of obtaining the value of $l^{(0)}$, which gives rise to an impurity level for given k and l. (For the case $A(k) < -1$.)

For negative energies we have to replace k by $\varkappa$. Then the function $A(\varkappa)$ either indefinitely increases or decreases for large l, according as $V/2\varkappa < 1$ or > 1. The former case corresponds to the zeroth gap, while the latter to the first gap for $V > 0$. This can be seen from Fig. 4.15, in which the function $A(\varkappa)$ is plotted for several values of l. In place of (4.5.5), we have the equation

$$\exp\{2\varkappa(l^{(0)} - l)\} = 1 + \frac{8\varkappa^2}{V^2}\left\{(A^2(\varkappa) - 1) \pm C'(\varkappa)\,(A^2(\varkappa) - 1)^{1/2}\right\}, \tag{4.5.7}$$

where

$$C'(\varkappa) \equiv \sinh \varkappa l - (V/2\varkappa) \cosh \varkappa l, \tag{4.5.8}$$

and we should adopt the plus or minus sign, according as $A(\varkappa) < -1$ or $A(\varkappa) > 1$. When $V < 0$, both $C'(\varkappa)$ and $A(\varkappa)$ are positive, so that the right-hand side of (4.5.7) is smaller than unity. Hence in this case $l^{(0)}$ must be smaller than l. For $V > 0$ and $V/2\varkappa > 1$, both $C'(\varkappa)$ and $A(\varkappa)$ are negative, so that the right-hand side of

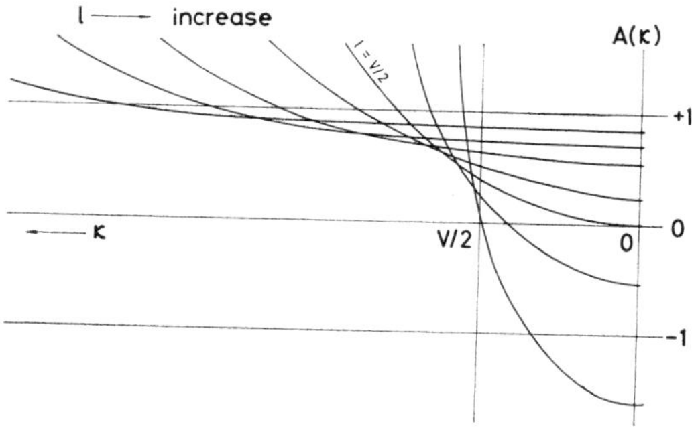

FIG. 4.15. The function $A(\varkappa)$ for various values of l (schematic).

(4.5.7) is less than unity, and we have $l^{(0)} < l$. Finally, for $V > 0$ and $V/2\varkappa < 1$, both $C'(\varkappa)$ and $A(\varkappa)$ are positive, and again $l^{(0)}$ must be smaller than l. In any case, the value of $l^{(0)}$ is uniquely determined, in contrast to the case of positive energy.

The variation of the phase in the neighbourhood of an impurity level is entirely similar to the foregoing cases, and need not be discussed again.

CHAPTER 5

Spectral Properties of Disordered One-dimensional Systems

5.1. Gaps in the Eigenfrequency Spectrum of a Polyatomic Chain

It was proved by Hori (1964a) that the Saxon–Hutner-type proposition is valid for several one-dimensional systems. In this section we first consider an isotopically disordered polyatomic chain. The masses of isotopes are denoted by $m^{(i)}$, $i = 0, 1, 2, \ldots, S-1$, in the order of increasing mass, and a mass $m^{(b)}$, which is smaller than $m^{(0)}$, is chosen as the standard mass. Further, we use the complex transfer matrices given by (3.2.8).

The eigenvalue equation and the eigenvalues of the transfer matrix $\mathbf{Q}^{(i)}$ associated with the atom of mass $m^{(i)}$ are

$$\theta^2 - 2A^{(i)}(\beta)\,\theta + 1 = 0 \tag{5.1.1}$$

and

$$\theta_{\pm}^{(i)} = A^{(i)}(\beta) \pm \{A^{(i)2}(\beta) - 1\}^{1/2} \tag{5.1.2}$$

respectively, where

$$A^{(i)}(\beta) \equiv \cos 2\beta - \tan \gamma^{(i)} \sin 2\beta\,. \tag{5.1.3}$$

The fixed points $z_{\pm}^{(i)}$ of the matrix $\mathbf{Q}^{(i)}$ are easily seen to be given by

$$z_{\pm}^{(i)} = \{(1 - i \tan \gamma^{(i)}) \exp(-2i\beta) - \theta_{\pm}^{(i)}\}/i \tan \gamma^{(i)} \exp(2i\beta)\,. \tag{5.1.4}$$

The matrix $\mathbf{Q}^{(i)}$ is hyperbolic, and consequently $z_{\pm}^{(i)}$ become limit points, when

$$A^{(i)}(\beta) < -1\,. \tag{5.1.5}$$

We have taken into account the fact that $A^{(i)}(\beta)$ is always negative for real $\theta_{\pm}^{(i)}$ in order to give the double sign in (5.1.2) the right order, so that $\theta_{-}^{(i)}$ may be given the larger modulus. From (5.1.3) and (5.1.5) it is seen that if, for a given value of β, the mass parameter $Q^{(0)}$ is larger than the critical value $Q_c(\beta)$ defined by

$$Q_c(\beta) = (\cos 2\beta + 1)/\tan \beta \sin 2\beta = \cot^2 \beta, \qquad (5.1.6)$$

the eigenvalues $\theta_{\pm}^{(i)}$ are real and distinct for every i. The critical value $Q_c(\beta)$ gives the mass $m_c = (1 + Q_c(\beta))\, m^{(b)}$, such that the maximum frequency of the regular lattice of the atom with this mass just corresponds to the given value of β. Every regular lattice with the mass larger than m_c has a smaller maximum frequency.

From (5.1.4) we easily obtain for the limit phases $\delta_{\pm}^{(i)}$ the formula

$$\cos (\delta_{\pm}^{(i)}/2) = \left\{\frac{\sin 2\beta}{-2\theta_{\pm}^{(i)} \tan \gamma^{(i)}}\right\}^{1/2}. \qquad (5.1.7)$$

Suppose temporarily that, for a given β, the value of $Q^{(i)}$ is distributed continuously from $Q_c(\beta)$ to infinity, and regard the transfer matrix, its eigenvalues, limit phases, etc., as functions of the continuous variable Q, the index i being omitted from all these quantities. It is evident that $\mathbf{Q}$ is hyperbolic throughout such a range of the parameter Q. For $Q = Q_c$, we have $\theta_{\pm} = -1$, and

$$\cos (\delta_{\pm}/2) = \sin \beta, \quad \text{or} \quad \delta_{\pm} = \pi - 2\beta < \pi. \qquad (5.1.8)$$

It is seen from (5.1.2) that as Q increases, θ_{-} decreases down to $-\infty$, while θ_{+} increases up to zero. To see how the limit phases vary with the increase of Q, we differentiate the square of (5.1.7) with respect to Q, obtaining

$$\frac{d\{(\theta_{\pm}^{-1} \sin 2\beta)/(-2 \tan \gamma)\}}{dQ} = \cos^2 \beta \, \frac{\{\theta_{\pm}^{-1} - Q d\theta_{\pm}^{-1}/dQ\}}{Q^2}. \qquad (5.1.9)$$

For the lower sign, this is always negative, while for the upper sign, this is at first positive, becomes zero at

$$\{(\cos 2\beta - Q \tan \beta \sin 2\beta)^2 - 1\}^{1/2} = \tan \gamma \sin 2\beta, \qquad (5.1.10)$$

and then becomes negative. This means that during the increase of Q from Q_c to infinity, δ_{-} increases up to $+\pi$, while δ_{+} decreases down to $\pi - 4\beta$. Thus the interval spanned by δ_{-} and that spanned by δ_{+} are disjoint, and, *a fortiori*, the sink- and source-phase intervals $\{\delta_-^{(i)}\}$ and $\{\delta_+^{(i)}\}$, $i = 0, 1, 2, \ldots, S-1$, are disjoint. Consequently, the proposition of the Saxon–Hutner-type is valid for the polyatomic isotopic lattice. Namely, if for an interval of β-values all the $Q^{(i)}$'s are larger than Q_c, all the matrices $\mathbf{Q}^{(i)}$ are hyperbolic, so that this interval lies in a spectral gap for every constituent lattice. Then the intervals $\{\delta_-^{(i)}\}$ and $\{\delta_+^{(i)}\}$ are disjoint throughout the interval of β, so that this interval lies in a spectral gap of the mixed lattice according to the theorem of Hori and Matsuda.

The above result is, however, rather trivial, for in the above choice of the set $\{\mathbf{Q}^{(i)}\}$, every constituent lattice is a monatomic one composed of each isotope, and its spectral gap is merely the frequency region which extends from the maximum frequency to infinity. Any interval of β which has the property mentioned above should necessarily lie in the frequency region higher than the maximum frequency of the regular lattice of the lightest atom with mass $m^{(0)}$. Hence the above result is merely another way of expressing the well-known fact that there can be no eigenfrequency larger than the maximum frequency of the regular lattice of the lightest atom.

Next consider a non-isotopic polyatomic chain. In this case it is convenient to use the real representation (1.3.10):

$$\mathbf{T}^{(i,j)} = \begin{pmatrix} 1 & 1 \\ -g^{(i)}\omega^2 & h^{(j)} - g^{(i)}\omega^2 \end{pmatrix}, \tag{5.1.11}$$

where $g^{(i)}$ and $h^{(j)}$ are respectively the ith and jth of the possible values of $m_n/K_{n,n+1}$ and $K_{n,n-1}/K_{n,n+1}$. Here each pair of indices (i, j) specifies a constituent regular lattice.

Suppose again that the quantities $g^{(i)}$ and $h^{(j)}$ are distributed continuously, and omit the indices i and j. The eigenvalues and the limit points of $\mathbf{T}$ are

$$\theta_\pm = [1 + h - g\omega^2 \pm \{(1 + h - g\omega^2)^2 - 4h\}^{1/2}]/2 \tag{5.1.12}$$

and

$$u_\pm = 1/(\theta_\pm - 1) \tag{5.1.13}$$

respectively. $\mathbf{T}$ is hyperbolic when either

$$\omega^2 \leqq (1 + h - 2h^{1/2})/g \tag{5.1.14}$$

or

$$\omega^2 \geqq (1 + h + 2h^{1/2})/g. \tag{5.1.15}$$

The former case is of no interest, for the right-hand side of (5.1.14) vanishes for $h = 1$. This value of h always appears in any lattice which is physically significant, and the frequency which satisfies the condition (5.1.14) for any value of g and h necessarily becomes imaginary. In the latter case we have $1 + h - g\omega^2 \leqq -2h^{1/2} < 0$, so that $\theta_\pm < 0$ and $1 - h + g\omega^2 > 2 > 0$. It is to be noted further that $(1 - h + g\omega^2)^2 > (1 + h - g\omega^2)^2 - 4h$.

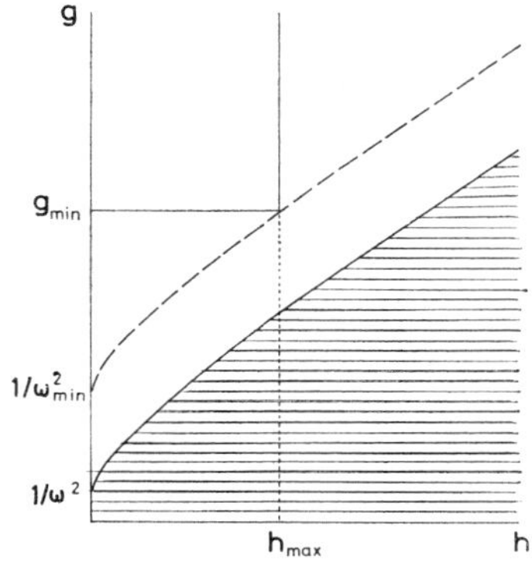

FIG. 5.1. The hyperbolic region (non-hatched) for the disordered polyatomic chain for given ω^2. In order to determine the minimum value of ω^2 for which **T** is hyperbolic for every pair (h, g), one has to choose ω_{min} such that the curve $1 + h + 2h^{1/2} = g\omega^2_{min}$ (broken curve) passes the point (h_{max}, g_{min}).

For a given value of ω^2, **T** is hyperbolic in the region shown in Fig. 5.1 (*hyperbolic region*). As ω^2 increases, the region grows wider. For a given range of values of g and h, the minimum value of ω^2 for which **T** is hyperbolic for every pair of values (h, g) is determined as shown in the same figure. On the curve $1 + h + 2h^{1/2} = g\omega^2$, $\theta_\pm$ and $u_\pm$ are equal to $(1 + h - g\omega^2)^2/2$ and $-1/(1 + h^{1/2})$ respectively. On the g-axis, i.e. for $h = 0$, we have $\theta_+ = 0$, $\theta_- = 1 - g\omega^2$, $u_+ = -1$, and $u_- = -1/g\omega^2$. For $g \to \infty$, we have $\theta_+ = 0$, $\theta_- = -\infty$, $u_+ = -1$, and $u_- = 0$.

In order to see whether the sets $\{u_-\}$ and $\{u_+\}$ are disjoint or not, we differentiate (5.1.12) with respect to g and h, obtaining

$$\frac{d\theta_\mp}{dg} = -\frac{\omega^2}{2}\left[1 \mp \frac{1 + h - g\omega^2}{\{(1 + h - g\omega^2)^2 - 4h\}^{1/2}}\right] \tag{5.1.16}$$

and

$$\frac{d\theta_\pm}{dh} = \frac{1}{2}\left[1 \mp \frac{1 - h + g\omega^2}{\{(1 + h - g\omega^2)^2 - 4h\}^{1/2}}\right] \tag{5.1.17}$$

respectively. By taking into account the conditions mentioned just above, it is seen that the right-hand sides of (5.1.16) and (5.1.17) are negative for the upper sign and positive for the lower sign. Hence θ_+ increases from $(1 + h - g\omega^2)/2$ up to zero either as h decreases or g increases, while θ_- decreases from $(1 + h - g\omega^2)/2$ down to $1 - g\omega^2$ as h decreases, and down to $-\infty$ as g increases. Then it is seen from (5.1.13) that u_+ decreases from $-1/(1 + h^{1/2})$ down to -1 either as h decreases or g increases, while u_- increases from $-1/(1 + h^{1/2})$ up to $-1/g\omega^2$ as h decreases and to zero as g increases. This situation is illustrated in Fig. 5.2.

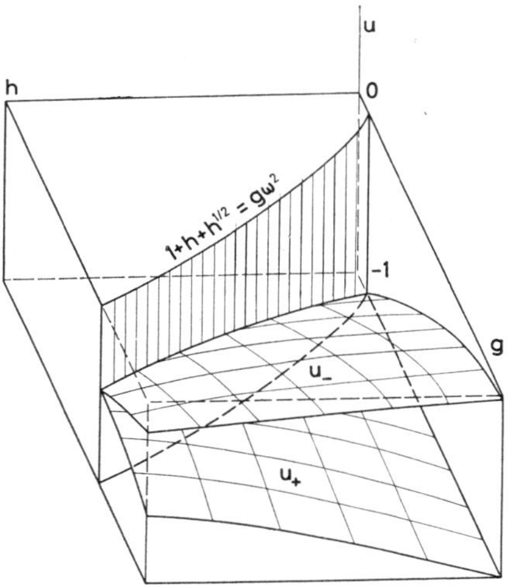

FIG. 5.2. The variation of limit points $u_\pm$ with g and h.

From the above result it is clear that the sets $\{u_-\}$ and $\{u_+\}$ are disjoint in any strip bounded by lines $g = g_{min}$, $h = h_{max}$, and $h = 0$ and lying wholly in the hyperbolic region, so that *a fortiori* the sets $\{u_-^{(i,j)}\}$ and $\{u_+^{(i,j)}\}$ are disjoint. Thus the proposition of the Saxon–Hutner type is valid also for non-isotopic polyatomic lattices. It should be noted, however, that in the present case the transfer matrix (5.1.11) is non-unimodular, and consequently the corresponding constituent lattice is non-physical, except the case $h^{(j)} = 1$. For such a non-physical lattice the word "spectral gap" has no real significance, but merely means the region of ω^2 for which $\mathbf{T}^{(i,j)}$ is hyperbolic.

Finally, consider a more complex case of glass-like diatomic chain mentioned in § 2.4. The transfer matrix associated with a unit cell is given by (1.6.15), where K_0 and K_1 vary randomly with some continuous distribution function. Its eigenvalues are given by

$$\theta_{\pm} = 1 - \frac{\lambda(K_0 + K_1)(m^{(0)} + m^{(1)}) - \lambda^2 m^{(0)} m^{(1)}}{2K_0 K_1} \pm \left[\left\{ 1 - \frac{\lambda(K_0 + K_1)(m^{(0)} + m^{(1)}) - \lambda^2 m^{(0)} m^{(1)}}{2K_0 K_1} \right\}^2 - 1 \right]^{1/2}, \tag{5.1.18}$$

which are real when

$$4K_0 K_1 \leqq \lambda(K_0 + K_1)(m^{(0)} + m^{(1)}) - \lambda^2 m^{(0)} m^{(1)} \tag{5.1.19a}$$

or

$$0 \geqq \lambda(K_0 + K_1)(m^{(0)} + m^{(1)}) - \lambda^2 m^{(0)} m^{(1)}. \tag{5.1.19b}$$

The hyperbolic region for fixed λ is shown in Fig. 5.3. The limit points $u_{\pm}$ are given by

$$u_{\pm} = -K_0 \times \frac{\theta_{\pm} + (K_0/K_1) - [(K_0 + K_1 - \lambda m^{(0)})(K_0 + K_1 - \lambda m^{(1)})]/K_0 K_1}{K_0 + K_1 - \lambda m^{(1)}} = \frac{K_0 + K_1 - \lambda m^{(0)}}{K_0\{\theta_{\pm} + (K_1/K_0)\}}. \tag{5.1.20}$$

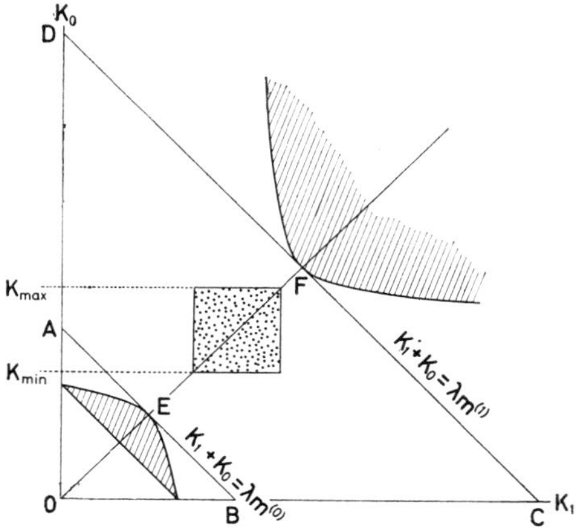

FIG. 5.3. The hyperbolic region (non-hatched) for the glass-like diatomic chain. The values of K_1 and K_0 are distributed in the dotted region.

For the lines *AEB* and *DFC* on which $K_0 + K_1 = \lambda m^{(0)}$ and $K_0 + K_1 = \lambda\, m^{(1)}$ respectively we have from (5.1.18)

$$\left.\begin{aligned} \theta_+ &= -K_1/K_0, \\ \theta_- &= -K_0/K_1, \end{aligned}\right\} \quad \text{on } AE \text{ and } DF,$$

$$\left.\begin{aligned} \theta_+ &= -K_0/K_1, \\ \theta_- &= -K_1/K_0, \end{aligned}\right\} \quad \text{on } EB \text{ and } FC. \tag{5.1.21}$$

This, together with (5.1.20), leads to the result

$$u_+ > 0, \quad u_- < 0, \quad \text{in the region } ABCD, \tag{5.1.22}$$

which shows that if K_0 and K_1 are distributed within the region *ABCD*, for example in the dotted region in Fig. 5.3, the sink and source intervals $\{u_-\}$ and $\{u_+\}$ are disjoint. Thus the proposition of Saxon–Hutner-type is valid, so that the interval of λ in which K_0 and K_1 are distributed within the region *ABCD* gives a gap of the given glass-like chain.

The gap must persist, therefore, between the top of the acoustical band of the constituent lattice with $K_1 = K_0 = K_{\max}$ and the bottom of the optical band of the lattice with $K_1 = K_0 = K_{\min}$, that is, between $\omega^2 = 2K_{\max}/m^{(1)}$ and $\omega^2 = 2K_{\min}/m^{(0)}$. In the case of Figs. 2.10a and 2.11a, b, the corresponding intervals of $m^{(0)}\omega^2$ are just those mentioned in § 2.4, i.e. (5/4, 3/2), (8/5, 3/2), and (3/4, 1) respectively. The gaps appearing in the figures are somewhat wider than these intervals. This is not a contradiction, since the proposition of the Saxon–Hutner type gives only a sufficient condition for the gaps, so that there remains the possibility that the density vanishes also outside these intervals. Such a widening of the gap might, however, be due to the inherent fault of the numerical calculation, which cannot detect too small values of density.

5.2. Gaps in the Energy Spectrum of a Disordered Kronig–Penney Model

Let us next turn to the consideration of a disordered Kronig–Penney model in which both the spacing of neighbouring potentials and the strength of the potential may vary randomly from cell to cell. Let us use the representation (1.3.20), and denote the transfer matrix associated with the cell which has the length $l^{(j)}$ and contains the potential of strength $-V^{(i)}$ by $\mathbf{T}^{(i,j)}$:

$$\mathbf{T}^{(i,j)} = \begin{pmatrix} \cos kl^{(j)} & \sin kl^{(j)}/k \\ -k \sin kl^{(j)} - V^{(i)} \cos kl^{(j)} & \cos kl^{(j)} - (V^{(i)}/k) \sin kl^{(j)} \end{pmatrix}. \tag{5.2.1}$$

The eigenvalue equation and eigenvalues of $\mathbf{T}^{(i,j)}$ are

$$\theta^{(i,j)2} - 2\theta^{(i,j)}A^{(i,j)}(k) + 1 = 0 \tag{5.2.2}$$

and

$$\theta_{+}^{(i,j)} = A^{(i,j)}(k) \pm (A^{(i,j)2}(k) - 1)^{1/2},$$

$$\theta_{-}^{(i,j)} = A^{(i,j)}(k) \mp (A^{(i,j)2}(k) - 1)^{1/2}, \tag{5.2.3}$$

respectively, where $A^{(i,j)} \equiv \cos kl^{(j)} - (V^{(i)}/2k) \sin kl^{(j)}$. The condition for $\mathbf{T}^{(i,j)}$ to be hyperbolic is that $|A^{(i,j)}(k)| > 1$. In contrast to the vibrational case, in which the corresponding quantity $A^{(i)}(\beta)$ cannot be larger than unity, both possibilities $A^{(i,j)}(k) > 1$ and $A^{(i,j)}(k) < -1$ may be realized. The odd gaps of the constituent lattice associated with $\mathbf{T}^{(i,j)}$ correspond to the case $A^{(i,j)}(k) < -1$ and the even gaps to $A^{(i,j)}(k) > 1$. The eigenvalues $\theta_{\pm}^{(i,j)}$ are negative or positive according as $A^{(i,j)}(k) < -1$ or $> +1$, so that we should adopt the upper sign when $A^{(i,j)}(k) < -1$ and the lower sign when $A^{(i,j)}(k) > +1$, in (5.2.3). The sign of $\theta_{\pm}^{(i,j)}$ is opposite to that of $\sin kl^{(j)}$ for $V^{(i)} > 0$, while for $V^{(i)} < 0$ these signs agree with each other.

The limit points are given by

$$\begin{aligned} u_{\pm}^{(i,j)} &= \frac{\cos kl^{(j)} - (V^{(i)}/k) \sin kl^{(j)} - \theta_{\pm}^{(i,j)}}{k \sin kl^{(j)} + V^{(i)} \cos kl^{(j)}} \\ &= \frac{-(V^{(i)}/2k) \sin kl^{(j)} \mp (A^{(i,j)2}(k) - 1)^{1/2}}{B^{(i,j)}(k)}, \quad A^{(i,j)}(k) < -1, \\ &= \frac{-(V^{(i)}/2k) \sin kl^{(j)} \pm (A^{(i,j)2}(k) - 1)^{1/2}}{B^{(i,j)}(k)}, \quad A^{(i,j)}(k) > +1, \end{aligned} \tag{5.2.4}$$

where $B^{(i,j)}(k) \equiv k \sin kl^{(j)} + V^{(i)} \cos kl^{(j)}$.

In the following considerations we regard, as in the preceding section, the quantities $\theta_{\pm}^{(i,j)}$, $u_{\pm}^{(i,j)}$, $A^{(i,j)}$, etc., as functions of the continuous variables l and V, omitting the suffices i and j everywhere and regarding k as a fixed parameter.

At first let us consider how $u_\pm$ change with l. In Fig. 5.4 the function $A(k)$ is plotted as a function of l, for fixed k and V. The curves are similar to those in Fig. 3.5, where $A(k)$ is plotted as a function of k for fixed V and l. Let us call the intervals of l in which $|A(k)| > 1$ the zeroth, first, second, ..., hyperbolic intervals in the order of increasing l, but omit the zeroth in the case of $V > 0$. Further, we call the first, third, ..., intervals odd, and the second, fourth, ..., intervals even, as in § 3.4.

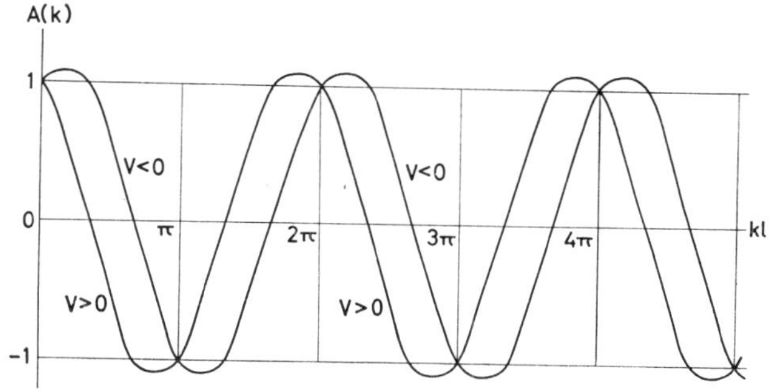

FIG. 5.4. The function $A(k)$, plotted as a function of l.

At $l = n\pi/k$, where n is a positive integer, both u_+ and u_- vanish:

$$u_+ = u_- = 0, \tag{5.2.5}$$

while at

$$A(k) = \pm 1 \tag{5.2.6}$$

they become

$$u_\pm = -(V/2k) \sin kl/B(k). \tag{5.2.7}$$

The zeros of the function $B(k)$ lie in the hyperbolic interval, for if $B(k) = 0$, $A^2(k)$ becomes

$$A^2(k) = 1 + (V^2/4k^2) \sin^2 kl, \tag{5.2.8}$$

which is larger than unity. Conversely, in every hyperbolic interval there is one zero of $B(k)$. This means that the values of $B(k)$ at the boundaries of each interval have opposite signs. Since at $l = n\pi/k$ $B(k)$ is equal to $-V$ or V according as n is odd or even, these signs are as shown in Table 4.1, provided the spectral gaps are replaced by corresponding hyperbolic intervals. From (5.2.7), the signs of $u_\pm$ at the boundaries can be obtained as given in the last line of the table.

At $B(k) = 0$, the denominator of (5.2.4) vanishes, so that here either u_+ or u_- must pass the infinity. By an argument entirely similar to that given in § 4.4, it can be shown that the limit point which becomes infinity at $B(k) = 0$ is u_+ and not u_-.

Applying the transformation $z = (u - i)/(u + i)$, the real axis is mapped on to the unit circle of the complex plane as shown in Fig. 4.9a. Hence the variation of the limit phases with the increase of l in each hyperbolic interval is as shown in Fig. 4.9b, c, where in the present case the spectral gaps are to be replaced by corresponding hyperbolic intervals.

For negative energies, we have to put $k = i\varkappa$, which amounts to merely replacing in the above formulae k by $\varkappa$ and the trigonometric functions by corresponding hyperbolic ones. The curve of $A(\varkappa)$ is shown in Fig. 5.5a, b, which correspond to the cases $V/2\varkappa < 1$ and $V/2\varkappa > 1$ respectively. In the case of $V < 0$, we have $A(\varkappa) = 1$ and $u_\pm = 0$ at $l = 0$. It is easily seen that for $l \to \infty$, $u_+ = -1/\varkappa$ (< 0) and $u_- = 1/(\varkappa - V) > 0$. Since $B(\varkappa)$ is always negative, we see from (5.2.4) that u_+ is always negative while u_- is always positive in this hyperbolic interval.

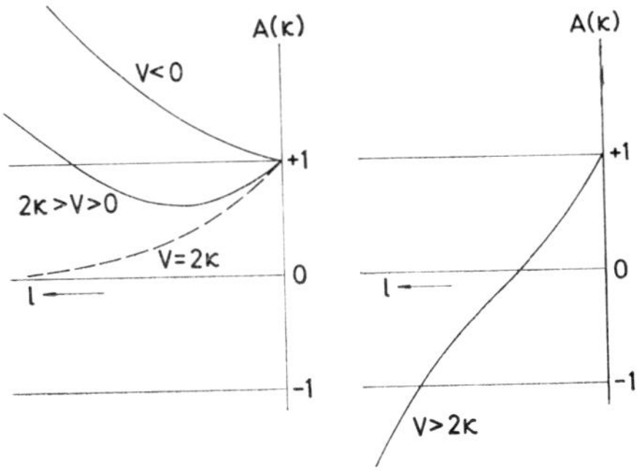

FIG. 5.5. The function $A(\varkappa)$ plotted as a function of l.

In the case where $V > 0$, $B(\varkappa)$ becomes $\frac{1}{2}(V - \varkappa)\exp(\varkappa l)$ for $l \to \infty$, so that $B(\varkappa)$ becomes positive or negative infinity according as $V > \varkappa$, or $V < \varkappa$. If $V < \varkappa$, $B(\varkappa)$ changes its sign in the hyperbolic interval, since then $B(\varkappa)$ becomes zero in this interval. Hence for $V < \varkappa$, $B(\varkappa)$ must be positive at $A(\varkappa) = 1$. For $V > \varkappa$ it is evident that $B(\varkappa)$ is always positive. According to (5.2.7) this means that $u_\pm$ must be negative at $A(\varkappa) = \pm 1$. The limiting values of $u_\pm$ for $l \to \infty$ is easily calculated from (5.2.4) to be $u_+ = -1/\varkappa$ and $u_- = 1/(\varkappa - V)$. For $V < \varkappa\,(A > +1)$, u_+ is negative while u_- is positive. For $2\varkappa > V > \varkappa\,(A > +1)$, $u_\pm$ are both negative and $u_- < u_+$. For $V > 2\varkappa\,(A < -1)$, $u_\pm$ are both negative and $u_- > u_+$.

Differentiating (5.2.4) with respect to l and examining the dominant term at $|A| = 1$, we see that du_+/dl is negative or positive according as $A = -1$ or $+1$, and du_-/dl is positive or negative according as $A = -1$ or $+1$.

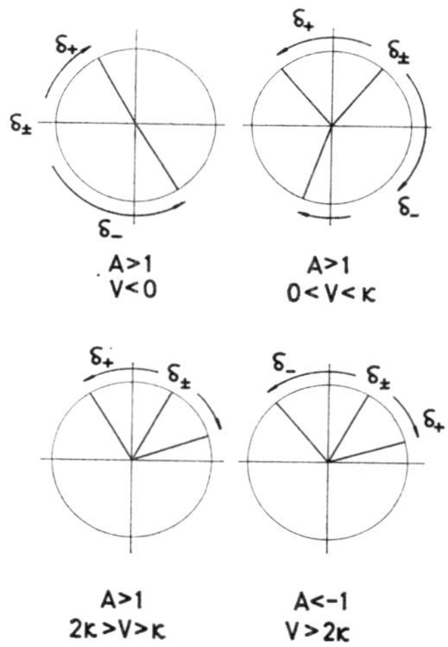

FIG. 5.6. The variation of limit phases with the increase of l for a negative energy.

From the above results it is concluded that the variation of phases with the increase of l is as shown in Fig. 5.6.

We should next examine how $u_\pm$ change with V. For a fixed k, the regions in the $(l - V)$ plane in which the transfer matrix is hyperbolic (hyperbolic region) are shown in Fig. 5.7, where the values and signs taken by $u_\pm$ on each boundary are indicated, together with the sign of $B(k)$ (see Table 4.1). The limiting values of u_+ and u_- for $|V| \to \infty$ for each fixed l are $-(1/k)\tan kl$ and zero respectively, and the limiting value of $u_\pm$ along the curved boundary is $-(1/2k)\tan kl$. It was shown above that on the curve $B(k) = 0$, u_+ passes infinity, whereas u_- varies continuously.

Differentiating (5.2.4), we get

$$\frac{du_\pm}{dV} = \frac{-\frac{1}{2}(A^2-1)^{1/2}\sin^2 kl \mp \{(A^2-1)\cos kl + (\sin kl/2k)AB\}}{B^2(A^2-1)^{1/2}}, \quad A > 1,$$

$$= \frac{-\frac{1}{2}(A^2-1)^{1/2}\sin^2 kl \pm \{(A^2-1)\cos kl + (\sin kl/2k)AB\}}{B^2(A^2-1)^{1/2}}, \quad A < -1, \qquad (5.2.9)$$

from which we see that du_+/dV is negative for $|A| \to 1$, while du_-/dV is positive for $|A| \to 1$, in all cases. It may further be shown that also for large $|V|$ du_+/dV is negative while du_-/dV is positive, and on the curve $B = 0$, the former becomes infinity while the latter remains finite.

From these results we obtain the conclusion that, in the case of $V > 0$, u_+ always decreases and u_- always increases with the increase of V, while in the case of $V < 0$, u_+ always increases and u_- always decreases with the increase of $-V$. Further, u_+ always passes infinity if $0 < kl < \pi/2 \pmod{\pi}$ for $V < 0$ and if $\pi/2 < kl < \pi \pmod{\pi}$ for $V > 0$, whereas u_- never becomes infinity. The manner of variation of $u_\pm$ as functions of l and V is indicated in Fig. 5.7 and illustrated in Fig. 5.8 for the case of $V < 0$ and $A > 0$.

It is clear that the sets $\{u_-\}$ and $\{u_+\}$ are disjoint in any strip bounded either by the lines $l = l_{\min}$, $V = V_{\min}$, and $l = n\pi/k$ (the

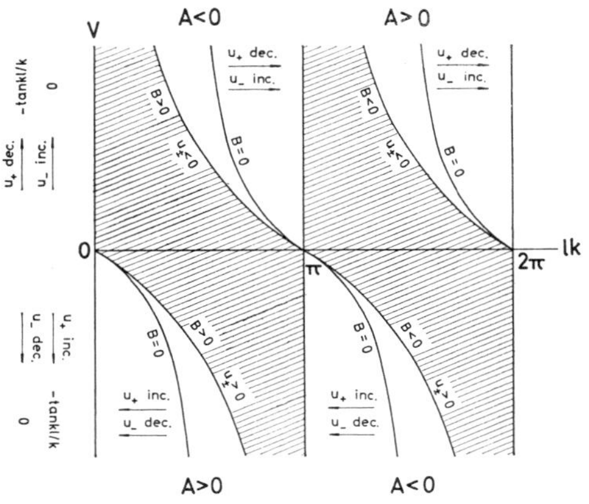

FIG. 5.7. The hyperbolic regions and the manner of variation of the limit points $u_\pm$ for the disordered Kronig–Penney model.

case of $V > 0$), or by the lines $l = l_{max}$, $V = V_{max}$ and $l = n\pi/k$ (the case of $V < 0$).

The argument runs in much the same way for the negative-energy region. Firstly, in the case of $V < 0$, it was shown above that u_+ is always negative while u_- is always positive. Consequently the sets $\{\delta_-\}$ and $\{\delta_+\}$ are necessarily disjoint for any distribution of positive potentials and the spacing between them.

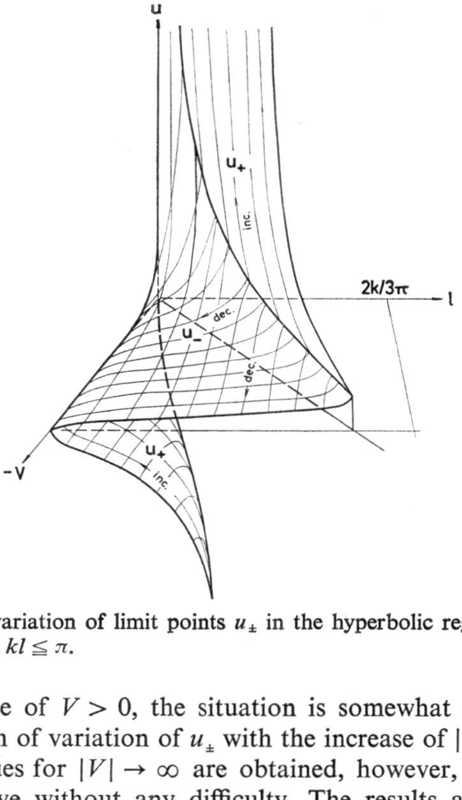

FIG. 5.8. The variation of limit points $u_\pm$ in the hyperbolic region for $V < 0$ and $0 \leqq kl \leqq \pi$.

In the case of $V > 0$, the situation is somewhat complicated. The direction of variation of $u_\pm$ with the increase of $|V|$, and their limiting values for $|V| \to \infty$ are obtained, however, in the same way as above without any difficulty. The results are shown in Fig. 5.9, together with the results previously obtained for the variation with l. It is seen that also here the disjointness condition is satisfied.

To sum up, we reach the conclusion that if the rectangle on $(l - V)$ plane defined by the inequalities $l_{min} \leqq l \leqq l_{max}$ and

$V_{min} \leqq V \leqq V_{max}$ is completely contained in a connected hyperbolic region in the upper or lower half-plane for some interval of k, this interval lies in a spectral gap of the mixed system.

The general result drawn above involves, of course, the well-known Saxon–Hutner theorem derived by Saxon and Hutner (1949) and Luttinger (1951) for the diatomic Kronig–Penney model. It states that the common spectral gap of two regular

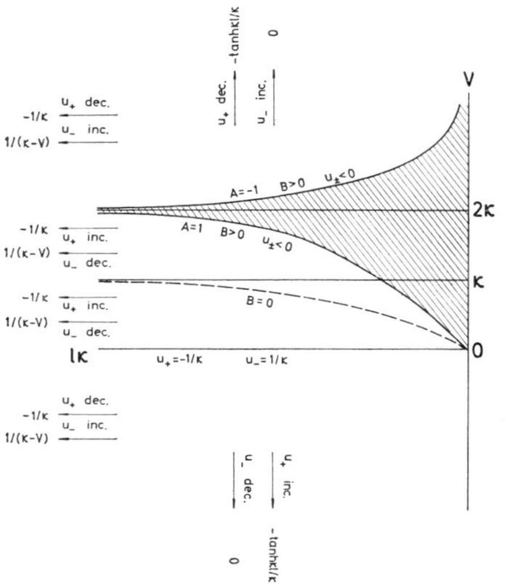

FIG. 5.9. The hyperbolic regions and the manner of variation of $u_{\pm}$ for a negative energy.

lattices, each of which is composed of one of the component atoms of mixed lattice, remains to be a gap also in the spectrum of the mixed system. For the diatomic chain considered by Agacy and Borland (1964), the range of ν from 1·504 to $\pi/2$ must therefore be a gap. The result of numerical calculation mentioned in § 2.5 is in harmony with this requirement, although the calculated gap seems to be somewhat wider than this interval. As was mentioned in the preceding section, this is not a contradiction. However, for the one-dimensional diatomic chain, Young and Dworin (1965) proved that any interval of k, which is allowed for any of the constituent atoms, is also allowed for the mixed lattice, provided that the number of repetitions of one and the same potential is not restricted to be finite. Thus in such a case only possible gaps in the mixed system are the intervals which are forbidden for every constituent system. In the present case, it can therefore be concluded that the apparent widening of the gap is due to the numerical nature of the calculation, which makes it difficult to detect too small densities.

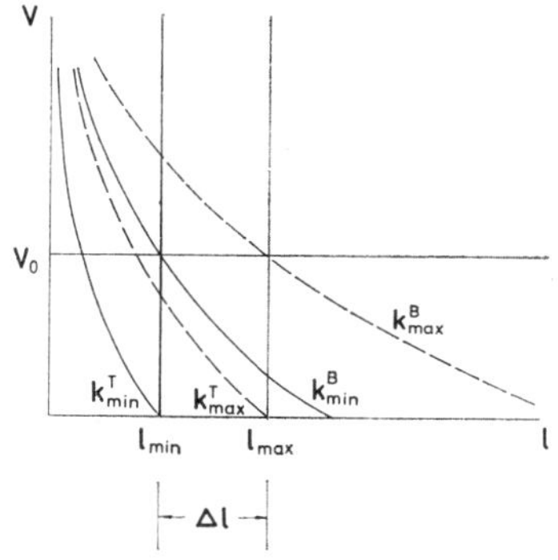

FIG. 5.10. The determination of gaps for the liquid-metal model.

Next consider the case of the liquid-metal model, in which V is fixed at V_0 while l varies randomly over an interval of width Δl. As an example, consider the case of $V_0 > 0$. For l_{min}, the boundary curves which pass the points (l_{min}, V_0) and $(l_{min}, 0)$ give two values of k, k^B_{min} and k^T_{min}, corresponding to the bottom and top respectively of a gap of the regular lattice characterized by V_0 and l_{min}. These curves are drawn in Fig. 5.10 by full lines. The broken lines and k^B_{max} and k^T_{max} are the corresponding curves and k-values for l_{max}. The interval of k, (k^B_{min}, k^T_{max}), should just correspond to the gap of the disordered lattice. If Δl increases, the width of this interval becomes smaller, and when Δl reaches the value Δl_{crit}, for which k^B_{min} and k^T_{max} coincide with each other, the gap vanishes and persists no longer. It is clear that the value of Δl_{crit} becomes smaller and smaller as k increases, since the boundary curves come out rapidly steeper. Therefore, the condition for the persistence of the gap becomes severer for higher gaps.

The values of k^B_{min} and k^T_{max} are given by

$$\cos k^B_{min} l_{min} - (V_0/2k^B_{min}) \sin k^B_{min} l_{min} = \pm 1, \qquad (5.2.10)$$

and

$$\cos k^T_{max} l_{max} = \pm 1 \qquad (5.2.11)$$

respectively. Equation (5.2.10) can be reduced to

$$n\pi - 2\tan^{-1}(V_0/2k^B_{\min}) = k^B_{\min} l_{\min}, \tag{5.2.12}$$

where n is an integer, while (5.2.11) is equivalent to

$$k^T_{\max} l_{\max} = n\pi, \quad \text{or} \quad k^T_{\max} l_{\min} = n\pi - k^T_{\max} \Delta l. \tag{5.2.13}$$

Equations (5.2.12) and (5.2.13) are simply those derived by Borland (1961 b) and Roberts and Makinson (1962). The integer n corresponds to the order number of the gap (see § 3.4).

In the negative-energy region, the extension of the gaps is determined, as can be seen from Fig. 4.15, by the value of the minimum spacing $l_{\min}$. The equation for the gap boundary is

$$\cosh \varkappa l_{\min} - (V_0/2\varkappa) \sinh \varkappa l_{\min} = \pm 1, \tag{5.2.14}$$

where the plus and minus signs correspond to the upper edge of the zeroth gap and the lower edge of the first gap respectively.

In the calculation of Makinson and Roberts (1960) and Roberts (1963) mentioned in § 2.5, the values of $l_{\max}$ were 1·0346 for the case of $\sigma = 0{\cdot}02$ and 1·0866 for $\sigma = 0{\cdot}05$. According to (5.2.13), this gives $k^T_{\max}$ the values 3·036 and 2·891 respectively. From Fig. 2.15 it is seen that for both cases we have $k^T_{\max} < k^B_{\min}$, so that the gap may not persist. By a brief calculation the same conclusion is attained also for the cases treated by Borland and Bird (1964) (Fig. 2.16). Thus the gaps obtained by numerical calculation may be only apparent. Again we cannot exclude, however, the possibility that they are true gaps. For example, the calculated gaps may be true ones which are to be predicted by some other proposition of the Saxon–Hutner type, corresponding to a different choice of the set of constituent lattices.

5.3. Special Frequencies in the Isotopically Disordered Chain

Returning again to the vibrational problem, consider a linear disordered chain composed of two isotopes with masses $m^{(0)}$ and $m^{(1)}$ ($m^{(0)} < m^{(1)}$). We assume at first that there appears no succession of more than $s - 1$ light atoms. We have seen already in § 3.5 that in such a case we can choose the generalized diatomic lattices with $p = 0, 1, 2, \ldots, s - 1$, as the constituent systems, and that, when the mass ratio reaches a certain value, there appear, just below one or some of the bands of these constituent lattices, narrow β-intervals each of which is a common gap for every constituent lattice. If the disjointness condition is fulfilled, these intervals must be gaps also for the mixed lattice. As $Q^{(1)}$ further increases, new gaps successively appear below other bands. Once appeared, their width continually increases with $Q^{(1)}$. For sufficiently large mass ratio, such gaps eventually appear below all the bands of the constituent lattices. Figures 5.11 and 5.12 show the manner of appearance of these common gaps.

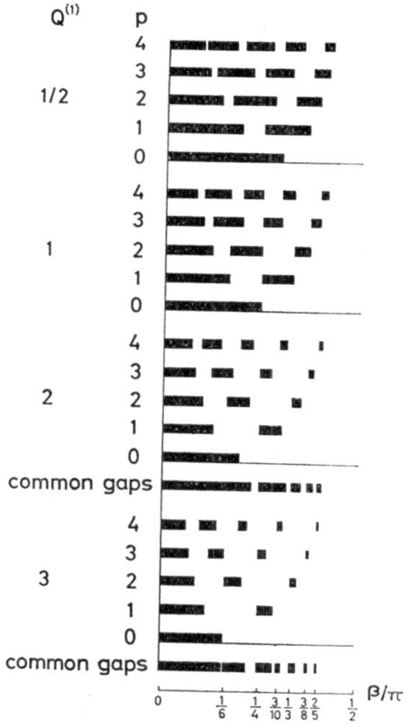

FIG. 5.11. The diminishing of the bands (black strips) of the constituent generalized diatomic lattices and the widening of common gaps, with the increase of $Q^{(1)}$.

As the number s is increased, the number of constituent lattices increases, and more and more bands participate to interlace one another. Hence it would seem at first sight that it becomes progressively more difficult for the common gaps to appear. In fact, however, this is not the case. The gaps that have already appeared remain to exist however large s becomes, although their width diminishes owing to the interlacing of the bands. Moreover, every band that newly appears is accompanied below its bottom by a gap, provided the mass ratio is sufficiently large. Finally, for $s \to \infty$ and infinite mass ratio, all the β-values which are rational multiples of π are associated with infinitesimal gaps of the disordered chain.

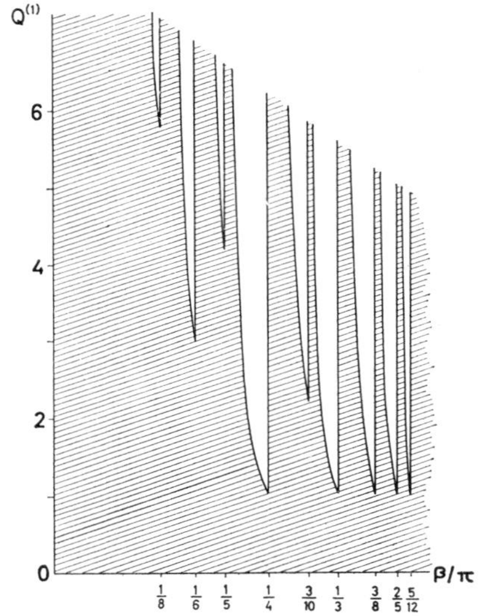

FIG. 5.12. The variation of the common gaps with the mass ratio, for $s = 6$. The non-hatched region corresponds to the common gaps.

The frequency with which such a gap is associated is called *special frequency*, following the terminology of Matsuda (1964) who first suggested and proved its existence. The phase-theoretical treatment of the special frequency was given for the first time by Hori (1964b, 1966). In order to confirm the above statements, it is sufficient to show that there exists, for each rational-multiple value of β, a critical mass parameter $Q_c(\beta)$, above which every matrix of the set $\{\mathbf{Q}^{(1p)}\}, p = 0, 1, 2, \ldots,$ is hyperbolic, and, moreover, the phase-disjointness condition is satisfied.

Every positive number which is a rational multiple of π and at the same time smaller than (or equal to) $\pi/2$ can be written as

$(s - t)\pi/2s$, where s and t are integers, and $s \geqq 1$, $t \geqq 0$, and $s > t$. Since at the light atom the phase increases by 4β, any succession of s light atoms may be neglected as long as we are considering such a value of β and investigating only the conditions of hyperbolicity and phase disjointness of the set $\{\mathbf{Q}^{(1p)}\}$, disregarding the *amount* of phase change. After omitting these successions from the lattice, we obtain a chain in which there is no succession of more than $s - 1$ light atoms. The remaining lattice (*reduced lattice*) is just that described by the finite set $\{\mathbf{Q}^{(1p)}\}$, $p = 0, 1, 2, \ldots, s - 1$.

The eigenvalue equation and the limit points of $\mathbf{Q}^{(1p)}$ are

$$\theta^2 - 2A^{(1)}((p + 1)\beta)\,\theta + 1 = 0 \tag{5.3.1}$$

and

$$z_{\pm}^{(1p)} = \frac{-2\theta_{\pm}^{(1p)} \exp(-2(p + 1)\, i\beta) + (\theta_{\pm}^{(1p)})^2 + \exp(-4(p + 1)\, i\beta)}{2\theta_{\pm}^{(1p)} \cos(2(p + 1)\beta) - (\theta_{\pm}^{(1p)})^2 - 1} \tag{5.3.2}$$

respectively, from which we obtain for the phases $\delta_{\pm}^{(1p)}$

$$\cos\left\{\frac{\delta_{\pm}^{(1p)}}{2}\right\} = \left\{\frac{\sin(2(p + 1)\beta)}{-2\theta_{\pm}^{(1p)} \tan\gamma^{(1)}}\right\}^{1/2}. \tag{5.3.3}$$

It is seen from (5.3.1) that $\theta_{\pm}^{(1p)}$ are positive or negative according as $\sin(2(p + 1)\beta)$ is negative or positive. Since, however, $\cos(\delta_{\pm}^{(1p)}/2)$ is invariant under the translation $2(p + 1)\beta \to 2(p + 1)\beta + \pi$, we may bring all the angles $\alpha \equiv 2(p + 1)\beta$, $p = 0, 1, 2, \ldots, s - 1$, into the interval $(0, \pi)$, by subtracting from them appropriate integral multiples of π, as far as we are considering the distribution of $\delta_{\pm}^{(1p)}$ for a given value of β.

The condition for the eigenvalues $\theta_{\pm}^{(1p)}$ to be real is then given by

$$A^{(1)}(\alpha) \equiv \cos\alpha - Q^{(1)} \tan\beta \sin\alpha \leqq -1. \tag{5.3.4}$$

This should give the spectral gaps of the constituent lattices. If we regard α as a continuous variable, and plot the function $A^{(1)}(\alpha)$ (Fig. 5.13), it is seen that, in case (5.3.4) is satisfied by the value of p, $p_{\min}$, which gives the smallest non-zero α-value, $\alpha_{\min}$, it is automatically satisfied by all the other values of p. The condition (5.3.4) is, on the other hand, fulfilled by $p_{\min}$ if $Q^{(1)}$ is larger than or equal to some critical value $Q_c(\beta)$. In other words, if $Q^{(1)} > Q_c(\beta)$, the β-value under consideration lies in one of the spectral gaps for every constituent lattice.

The value of α for which $\theta_{\pm}^{(1p)} = -1$ is given by

$$2 + 2(\cos\alpha - \tan\gamma^{(1)}\sin\alpha) = 0, \quad \text{or} \quad \sin\alpha = \sin 2\gamma^{(1)}. \tag{5.3.5}$$

By putting this into (5.3.3), it reduces to

$$\cos(\delta_{\pm}^{(1p)}/2) = 1/(Q^{(1)} + 1)^{1/2}. \tag{5.3.6}$$

For $\alpha = \pi$, $\cos(\delta_{\pm}^{(1p)}/2)$ becomes zero. Differentiating the square of the right-hand side of (5.3.3), we get

$$\frac{d\{-\sin\alpha/(2\theta_{\pm}^{(1p)}\tan\gamma^{(1)})\}}{d\alpha} = \frac{\{(\cos\alpha - \tan\gamma^{(1)}\sin\alpha)^2 - 1\}^{1/2} \mp \tan\gamma^{(1)}\sin\alpha}{-2\tan\gamma^{(1)}\{\cos\alpha - \tan\gamma^{(1)}\sin\alpha \pm (A^{(1)2} - 1)^{1/2}\}^2 (A^{(1)2} - 1)^{1/2}}. \tag{5.3.7}$$

For the lower sign, this expression is always negative, while for the upper sign, it is at first positive, vanishes at $(A^{(1)2} - 1)^{1/2} = \tan\gamma^{(1)}\sin\alpha$, when $\cos(\delta_{+}^{(1p)}/2) = 1$, and then becomes negative. This means that as α increases, $\delta_{-}^{(1p)}$ increases from the value given by (5.3.6) up to π, while $\delta_{+}^{(1p)}$ decreases down to $-\pi$, so that the source- and sink-phase intervals $\{\delta_{+}^{(1p)}\}$ and $\{\delta_{-}^{(1p)}\}$ are disjoint with each other.

The critical mass parameter $Q_c(\beta)$ can be easily obtained from (5.3.4). The value of $\alpha_{\min}$ is given by π/s, irrespective of whether s

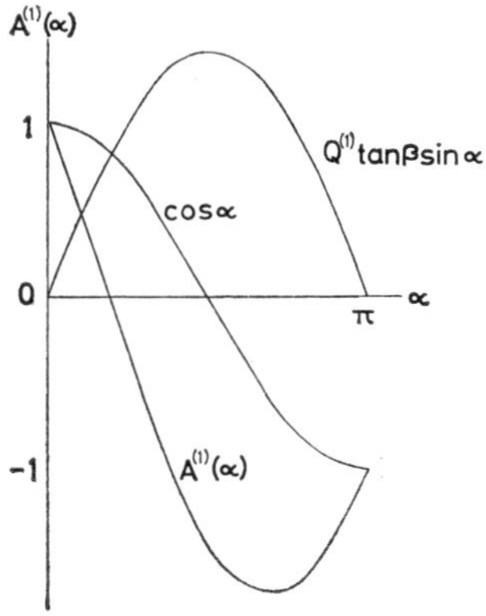

FIG. 5.13. The plot of the function $A^{(1)}(\alpha)$.

is even or odd, unless the form $(s - t)/2s$ is reducible to the same form with a smaller s. For $s > 1$, (5.3.4) gives

$$Q_c(\beta) = \frac{1 + \cos(\pi/s)}{\tan\{(s-t)\pi/2s\}\sin(\pi/s)} = \cot\left(\frac{\pi}{2s}\right)\tan\left(\frac{t\pi}{2s}\right). \qquad (5.3.8)$$

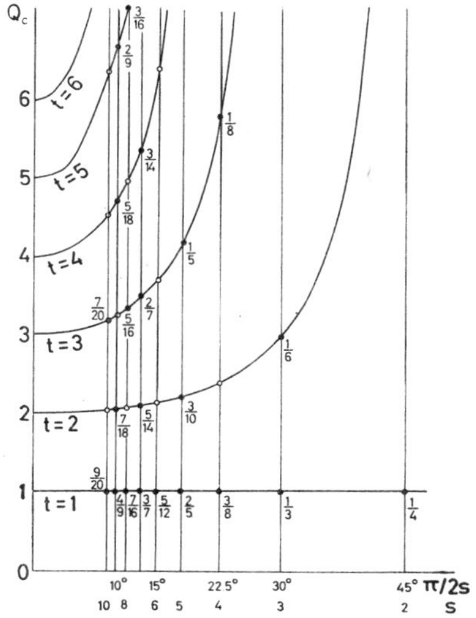

FIG. 5.14. The critical mass parameter Q_c as a function of s and t (up to $s = 10$). The value of Q_c at each pair (s, t) is indicated by a black circle. The fraction attached to each circle is the value of $(s - t)/2s$. White circles correspond to the reducible fractions and do not give new values of Q_c.

In Fig. 5.14, the values of $Q_c(\beta)$ are plotted as a function of s for each value of t. From this one can obtain $Q_c(\beta)$ for any rational number. Each white circle corresponds to a fraction $(s - t)/2s$ which is reducible to a fraction of the same form with a smaller s, and does not give any new values of $Q_c(\beta)$.

For $s = 1(\alpha_{\min} = \pi)$, (5.3.4) is satisfied by an arbitrary value of $Q^{(1)}$. Since $s = 1$ corresponds to the maximum frequency of the regular lattice of the lightest atom, this is a matter of course.

From the figure we also see which rational numbers give special frequencies for a given value of $Q^{(1)}$. For $Q^{(1)} < 1$, there exists no special frequency. At $Q^{(1)} = 1$, all the rational numbers of the form $(s-1)/2s$ suddenly begin to give special frequencies. Up to $Q^{(1)} = 2$, no new special frequency appears. When $Q^{(1)}$ exceeds 2, rational numbers of the form $(s-2)/2s$ (except the reducible ones) begin to give special frequencies from the side of larger s, until at $Q^{(1)} = 3$ the final member of them gives a special frequency. Numbers of the form $(s-3)/2s$ begin to give special frequencies at $Q^{(1)} = 3$, and stop giving them at $Q^{(1)} = 5{\cdot}83$. In general, for given t, $Q^{(1)}$ must be larger than t in order that the rational numbers of the form $(s-t)/2s$, except those reducible to the same form with a smaller s, may give special frequencies. After $Q^{(1)}$ exceeds t, these begin to give special frequencies from the side of larger s, except when they are reducible. When $Q^{(1)}$ becomes

$$Q^{(1)} = \cot^2\left\{\frac{\pi}{2(t+1)}\right\}, \tag{5.3.9}$$

these numbers stop giving special frequencies, since the smallest value of s for given t is $s = t+1$.

In Fig. 5.15 the manner of appearance of the special frequencies explained above is illustrated more vividly. The special frequencies are depicted by vertical lines. In the high-frequency region, they

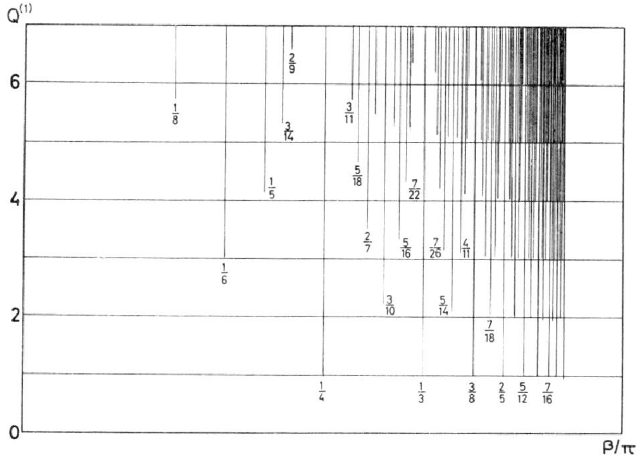

FIG. 5.15. The manner of appearance of the special frequencies, which are represented by vertical lines. The value of β/π for each special frequency is indicated at the starting point of the corresponding line. In the region $\beta/\pi > 9/20$, the lines become so dense that they cannot be depicted any more.

become denser and denser with such a rapidity that they cannot be drawn any more.

From the above argument it follows that the critical mass ratio is determined for each β of the form $(s - t)\pi/2s$ by considering only the associated reduced lattice, in which there appears no succession including more than $s - 1$ light atoms. This lattice is just the one considered at the beginning of this section, for which the last constituent lattice described by the matrix $\mathbf{Q}^{(1,\, s-1)}$ has the bands with their bottoms at just $\beta = (s - t)\pi/2s$. The situation that there appears a gap below the bottom of each band when $Q^{(1)}$ exceeds the critical value $Q_c(\beta)$ has thus been shown not to be affected by the presence of the sequence of light atoms longer than $s - 1$. On the contrary, the width of the gap is affected in such a way that it diminishes (for fixed mass ratio) when the sequences longer than $s - 1$ are added, since then there appear new constituent lattices with longer and longer unit cell, which brings about a rapid increase of the number of interlacing bands. These bands block up the space of unfolding of the already existing gaps, until finally the gaps have infinitesimal widths. In fact, the above proof of the hyperbolicity and phase disjointness is valid only at each special frequency.

Since, however, the probability of occurrence of the succession of light atoms becomes progressively small with the increase of its length, the density of the spectrum will remain small on the low-frequency side of the special frequency and vanish at that frequency itself. In contrast to this, the high-frequency side of the special frequency cannot belong to any gap from the outset, so that the density is not necessarily small on this side. It is expected generally that in an infinite lattice the spectral density jumps discontinuously from zero to a finite value at the special frequencies. If, accidentally, the mass parameter is exactly equal to the critical value $Q_c(\beta)$ corresponding to the value of β under consideration, the low-frequency side does not belong to a gap from the outset either, so that the density may not vanish also on this side. In other words, at the frequencies which have just become special, the spectrum may exhibit highly singular behaviour.

Notwithstanding such a possibility of singular behaviour, the spectral density in Dean's spectrum (Fig. 2.3) for $Q^{(1)} = 1$ seems to vanish well at $\beta = \pi/4, \pi/3, 3\pi/8, 2\pi/5, 5\pi/12, 3\pi/7, 4\pi/9, 7\pi/16, 9\pi/20$ and $15\pi/32$, i.e. just at the frequencies which were predicted above to become special at $Q^{(1)} = 1$. As Dean's spectrum has been obtained by numerical computation and is originally a histogram, it cannot be concluded from the figure that the density really varies in a continuous manner in the neighbourhood of the special frequencies. It may safely be concluded, however, that, at least in this example, the density vanishes or is extremely small at these frequencies, and that Dean's spectrum gives a model experimental proof of our theory.

From the origin of the special frequencies explained above, it is naturally expected that the integrated spectral density at any one of the special frequencies has a constant value independent of the mass parameter provided it is larger than the critical value, since no eigenfrequency can increase or decrease across the gap at the special frequency, however narrow it may be, during the increase of $Q^{(1)}$ from $Q_c(\beta)$ to infinity. Moreover, such an integrated density may easily be calculated analytically, for every rational number gives a special frequency in the limit $Q^{(1)} \to \infty$. (This is in accordance with the result obtained by Domb *et al.* (1959) that in this limit every rational number gives an eigenfrequency, for it has been claimed above that the infinitesimal interval *below* each special frequency is a gap of the disordered lattice.) Inasmuch as the lattice is decomposed in this limit into an infinite number of finite regular lattices of light atoms, the spectrum can be obtained simply by calculating the probability distribution of these finite lattices. Such a calculation was actually carried out by Matsuda and Teramoto (1965).

Finally, let us generalize the above result to the case of an isotopically disordered polyatomic chain. Let it be composed of S kinds of isotopes with masses $m^{(0)}, m^{(1)}, m^{(2)}, \ldots, m^{(S-1)}$ ($m^{(0)} < m^{(1)} < m^{(2)} < \ldots < m^{(S-1)}$), and choose the lightest atom as the basic one. For a given rational multiple value of β, the reduced lattice is composed of the following $(S-1)s$ kinds of groups of atoms:

$$\begin{array}{ll} m^{(1)}, m^{(0)}m^{(1)}, \ldots & \ldots, m^{(0)}m^{(0)} \ldots m^{(0)}m^{(1)} \\ m^{(2)}, m^{(0)}m^{(2)}, \ldots & \ldots, m^{(0)}m^{(0)} \ldots m^{(0)}m^{(2)} \\ \vdots \quad \vdots & \\ m^{(S-1)}, m^{(0)}m^{(S-1)}, \ldots & \ldots, \underbrace{m^{(0)}m^{(0)} \ldots m^{(0)}}_{s-1}m^{(S-1)} \end{array} \tag{5.3.10}$$

so that we have to consider $(S-1)s$ transfer matrices $\mathbf{Q}^{(ip)}$ ($i = 1, 2, \ldots, S-1$; $p = 0, 1, 2, \ldots, s-1$) and correspondingly

$(S - 1)s$ constituent lattices. It is clear from Fig. 5.13, however, that the eigenvalues $\theta_{\pm}^{(ip)}$ of $\mathbf{Q}^{(ip)}$ are real for all i and p, if only $Q^{(1)} > Q_c(\beta)$; in other words, if the given β lies in one of the spectral gaps for every constituent lattice with $i = 1$, it automatically lies in one of the gaps for any of the constituent lattices. Moreover, regarding $Q^{(1)}$ as a continuous variable Q and differentiating the square of the right-hand side of (5.3.3) with respect to it, we get

$$\frac{\theta_{\pm}^{-1} - Q\{d\theta_{\pm}^{-1}/dQ\}}{2Q^2 \tan\beta}, \tag{5.3.11}$$

for the derivative. As in the previous cases, it may be concluded from this formula that δ_- increases while δ_+ decreases with the increase of Q. Thus it is seen that for the β-value which satisfies $Q_c(\beta) < Q^{(1)}$, the source- and sink-phase intervals $\{\delta_+^{(ip)}\}$ and $\{\delta_-^{(ip)}\}$ ($i = 1, 2, \ldots, S - 1$ and $p = 0, 1, 2, \ldots, s - 1$) are disjoint. Therefore all the special frequencies remain to be special even if the atoms with the masses larger than $m^{(1)}$ are arbitrarily added.

It should be remarked that for the frequencies corresponding to $\beta = (s - 1)\pi/2s$, Borland also noted their special nature in 1964. He showed that at these frequencies the exact value of integrated spectral density may be calculated very easily, provided that $Q^{(1)}$, and consequently every $Q^{(i)}$ ($i = 2, \ldots, S - 1$), is larger than unity. Although his method of calculation differs from that used by Matsuda and Teramoto, his result is clearly a direct consequence of the above-mentioned characteristic property of special frequencies.

5.4. Special Energies in the Disordered Kronig–Penney Model

The existence of the *special points* at which the spectral density vanishes is not, of course, peculiar to the vibrational system. Wada (1966) showed that also in the disordered Kronig–Penney model there appear several *special energies* at which the density of the energy spectrum vanishes provided the difference between the strengths of potentials is sufficiently large.

Let us first consider the disordered diatomic Kronig–Penney model. Let the length of the cell be l, the strength of two potentials be $-V^{(0)}$ and $-V^{(1)}$, and denote the corresponding transfer matrices by $\mathbf{Q}^{(0)}$ and $\mathbf{Q}^{(1)}$, respectively. If we put

$$A^{(0)}(K) \equiv \cos kl - (V^{(0)}/2k) \sin kl \equiv \cos 2f, \qquad (5.4.1)$$

the eigenvalues of $\mathbf{Q}^{(0)}$ are easily seen to be given by $\exp(\pm 2if)$. Let us take the value of f in the interval $(0, \pi/2)$. It can be seen at once that, in the representation in which $\mathbf{Q}^{(0)}$ is diagonalized, every succession of s cells with the potential $-V^{(0)}$ may be neglected, provided that $f = t\pi/2s$, where t is an integer. Then we have only to consider the reduced lattice, which can be described by the set of matrices $\{\mathbf{Q}^{(1p)}\}$, $p = 0, 1, 2, \ldots, s-1$. Here the matrix $\mathbf{Q}^{(1p)} \equiv \mathbf{Q}^{(1)}\mathbf{Q}^{(0)p}$ is given by

$$\mathbf{Q}^{(1p)} = \begin{pmatrix} 1 + iR(f) & iR(f) \\ -iR(f) & 1 - iR(f) \end{pmatrix} \begin{pmatrix} \exp(2i(p+1)f) & 0 \\ 0 & \exp(-2i(p+1)f) \end{pmatrix}, \qquad (5.4.2)$$

where

$$R(f) \equiv (V^{(1)} - V^{(0)}) \sin kl/2k \sin 2f. \qquad (5.4.3)$$

By the same reason as in § 5.3, we may bring all the angles $\alpha \equiv 2(p+1)f$, $p = 0, 1, 2, \ldots, s-1$, into the interval $(0, \pi)$, by subtracting from them appropriate integral multiples of π. The condition for the value of f under consideration to lie in a spectral gap for every constituent lattice is that

$$|\text{trace}\, \mathbf{Q}^{(1p)}| > 2, \quad p = 0, 1, 2, \ldots, s-1, \qquad (5.4.4)$$

that is,

$$\cos(2(p+1)f) - R(f) \sin(2(p+1)f) < -1 \qquad (5.4.5\text{a})$$

or

$$\cos(2(p+1)f) - R(f) \sin(2(p+1)f) > +1. \qquad (5.4.5\text{b})$$

Equations (5.4.5a) and (5.4.5b) can be rewritten as

$$R(f) > \cot((p+1)f) \qquad (5.4.6\text{a})$$

and

$$R(f) < -\tan((p+1)f) \qquad (5.4.6\text{b})$$

respectively.

Suppose at first that $V^{(1)} > V^{(0)}$. Since $\sin 2f > 0$, $\cot[(p+1)f] > 0$ and $-\tan[(p+1)f] < 0$, the condition (5.4.6a) may be fulfilled only when $\sin kl > 0$, while (5.4.6b) may be satisfied only when $\sin kl < 0$.

The limit phases $\delta_{\pm}^{(1p)}$ are given by

$$\cos\left\{\frac{\delta_{\pm}^{(1p)}}{2}\right\} = \left\{\frac{\sin(2(p+1)f)}{-2\theta_{\pm}^{(1p)}R(f)}\right\}^{1/2}, \tag{5.4.7}$$

where $\theta_{\pm}^{(1p)}$ are the eigenvalues of $\mathbf{Q}^{(1p)}$. The condition (5.4.6a) is valid for all p if it is fulfilled by the minimum α, while (5.4.6b) is valid for all p if it is fulfilled by the maximum α. In both cases we can readily verify that the phase disjointness condition is satisfied by the set of matrices $\{\mathbf{Q}^{(1p)}\}$, by regarding α as a continuous variable and differentiating the square of the right-hand side of (5.4.7) with respect to α. Thus it is seen that each rational multiple value of f gives a special energy provided that $V^{(1)}$ is sufficiently larger than $V^{(0)}$.

The critical values of $V^{(1)}$, above which the value of f under consideration gives a special energy, can be obtained from (5.4.6a) and (5.4.6b) by putting $\alpha = \alpha_{\min} = \pi/s$ and $\alpha = \alpha_{\max} = (s-1)\pi/s$ respectively, and by replacing the inequality by equality:

$$V_c^{(1)} - V^{(0)} = \pm 2k \sin(t\pi/s)\cot(\pi/2s)/\sin kl. \tag{5.4.8}$$

Here the upper or lower sign must be used according as $\sin kl > 0$ or $\sin kl < 0$. The value of the right-hand side of (5.4.8) is plotted in Fig. 5.16 as a function of s and t for the interval of k which corresponds to the first band of the regular monatomic lattice with $V^{(0)}l = 0{\cdot}4$. The critical value of the potential $V_c^{(1)} - V^{(0)}$ at each pair (s, t) is indicated by a small black circle. The white circles correspond to reducible fractions t/s, and do not give new special energies. The three deep valleys at $\nu = 0{\cdot}23$, $\nu = 0{\cdot}41$, and $\nu = 0{\cdot}71$ in the spectrum calculated by Agacy and Borland (1964) (Fig. 2.12) are just at three special energies indicated in Fig. 5.16 by double circles.

It can easily be shown that the formula (5.4.8) is valid also for the case $V^{(1)} < V^{(0)}$, provided only that the plus sign is used when $\sin kl < 0$ and minus sign when $\sin kl > 0$.

From (5.4.6) it is expected that each rational multiple value of f gives the special energy even when other potentials $-V^{(i)}$, $i = 2, 3, \ldots$, are present, provided that $V^{(i)} \geqq V_c^{(1)}$ for all i in case $V_c^{(1)} > V^{(0)}$ and $V^{(i)} \leqq V_c^{(1)}$ for all i in case $V_c^{(1)} < V^{(0)}$. It is sufficient to show that the condition of phase disjointness is fulfilled by the set of matrices $\{\mathbf{Q}^{(ip)}\}$, $p = 0, 1, 2, \ldots, s-1$,

$i = 0, 1, 2, \ldots$ This can be proved by an entirely similar argument as that given in § 5.3 for the isotopically disordered polyatomic chain.

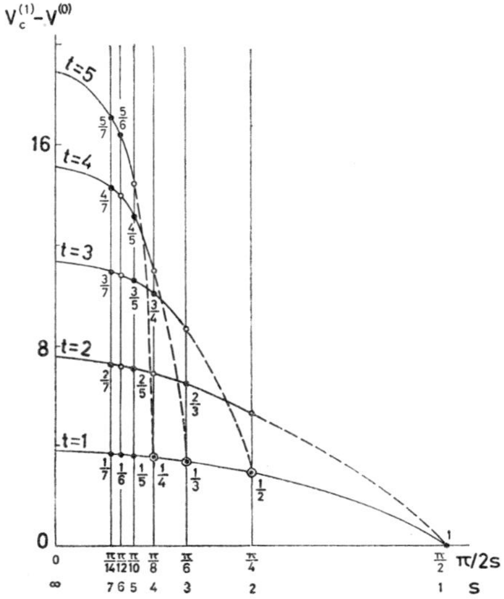

FIG. 5.16. The critical value $V_c^{(1)} - V^{(0)}$ for the interval of k which corresponds to the first band of the regular chain with $lV^{(0)} = 0{\cdot}4$, as a function of t and s. The value of $V_c^{(1)} - V^{(0)}$ at each pair (s, t) is indicated by a black circle. The fraction attached to each circle is the value of t/s. The white circles correspond to reducible fractions, and do not give new special energies. The double circles indicate the special energies appearing in the spectrum obtained by Agacy and Borland (Fig. 2.12).

5.5. Application of the Continued-fraction Theory to the Problem of Special Points

The existence of special points such as special frequencies and energies can also be proved by the continued-fraction theory of Matsuda and Okada. As an example, consider a linear chain for which the matrix $\mathbf{Q}_{1;N}$ can be represented as

$$\mathbf{Q}_{1;N} = \mathbf{Q}'\mathbf{Q}^{p(0)}\mathbf{Q}'\mathbf{Q}^{p(1)}\mathbf{Q}'\mathbf{Q}^{p(2)} \ldots \mathbf{Q}'\mathbf{Q}^{p(k)} \ldots \mathbf{Q}'\mathbf{Q}^{p(l)},$$
$$p(k) = 0, 1, 2, \ldots, s-1, \qquad (5.5.1)$$

where $\mathbf{Q}$ is a 2×2 unimodular matrix and $\mathbf{Q}'$ is an arbitrary 2×2 matrix with a real trace, both being continuous functions of λ. Let us use the representation in which $\mathbf{Q}$ is diagonal:

$$\mathbf{Q} = \begin{pmatrix} \exp(2i\beta) & 0 \\ 0 & \exp(-2i\beta) \end{pmatrix}, \tag{5.5.2}$$

and let the matrix $\mathbf{Q}'$ in this representation be

$$\mathbf{Q}' = \begin{pmatrix} a & b \\ c & d \end{pmatrix}. \tag{5.5.3}$$

If we identify the matrix in the formula (3.6.12) with $\mathbf{Q}'\mathbf{Q}^{p(k)}$, we have

$$\alpha_k = a\gamma^{2p(k)}, \quad \beta_k = b\gamma^{-2p(k)}, \quad \gamma_k = c\gamma^{2p(k)}, \quad \text{and} \quad \delta_k = d\gamma^{-2p(k)}, \tag{5.5.4}$$

where $\gamma \equiv \exp(2i\beta)$. Let us apply the theorem of Worpitzky to the expansion (3.6.15), where the quantities ϱ_k are $a_k b_k^{-1} b_{k-1}^{-1}$, which are now

$$\varrho_k = \frac{\gamma^{2\{p(k-2)-p(k-1)\}} \det \mathbf{Q}'}{\{a\gamma^{2p(k-1)} + d\gamma^{-2p(k-1)}\}\{a\gamma^{2p(k-2)} + d\gamma^{-2p(k-2)}\}},$$
$$k = 2, 3, \ldots, l. \tag{5.5.5}$$

If $|\text{trace}\, \mathbf{Q}| \leqq 2$ for $\lambda_1 < \lambda < \lambda_2$, β is real by (5.5.2). Then we have

$$|\varrho_k| = \frac{|\det \mathbf{Q}'|}{|\text{trace}\,(\mathbf{Q}'\mathbf{Q}^{p(k-1)})| \cdot |\text{trace}\,(\mathbf{Q}'\mathbf{Q}^{p(k-2)})|}. \tag{5.5.6}$$

Consequently, if

$$|\text{trace}\,(\mathbf{Q}'\mathbf{Q}^{p})| \geqq 2(|\det \mathbf{Q}'|)^{1/2}, \; p = 0, 1, 2, \ldots, s-1, \tag{5.5.7}$$

the inequalities (3.6.20) are satisfied, and the continued fraction $U(1, l)$ converges uniformly to U as $l \to \infty$.

When $\mathbf{Q}'$ is unimodular, the condition (5.5.7) is just the condition for $\mathbf{Q}'\mathbf{Q}^p$ to be hyperbolic or parabolic. Thus the above result means that if an interval (λ_1, λ_2) lies in the band of the regular lattice described by $\mathbf{Q}$, and at the same time in one of the gaps for every constituent lattice described by $\mathbf{Q}'\mathbf{Q}^p$, $p = 0, 1, 2, \ldots, s-1$, then the interval is in a gap of the mixed system described by $\mathbf{Q}_{1;N}$ given by (5.5.1).

In order to derive the result previously obtained for the isotopically disordered chain, consider a special case in which

$$\mathbf{C}_n = C - E \quad \text{or} \quad C' - E',$$
$$\mathbf{D}_n = D. \tag{5.5.8}$$

In this case we have

$$\mathbf{T}_n = \mathbf{T} \quad \text{or} \quad \mathbf{T}', \tag{5.5.9}$$

where

$$\mathbf{T} = \begin{pmatrix} (E - C)/D & -1 \\ 1 & 0 \end{pmatrix}, \quad \mathbf{T}' = \mathbf{T} + \begin{pmatrix} -\Delta C/D & 0 \\ 0 & 0 \end{pmatrix}, \tag{5.5.10}$$

and

$$\Delta C \equiv C' - C + (E - E'). \tag{5.5.11}$$

In the representation in which $\mathbf{T}$ is diagonalized to become $\mathbf{Q}$ as given by (5.5.2), the matrix $\mathbf{T}'$ comes out to be

$$\mathbf{Q}' = \mathbf{Q} - \frac{\Delta C/D}{\exp(2i\beta) - \exp(-2i\beta)} \begin{pmatrix} \exp(2i\beta) & 1 \\ -1 & -\exp(-2i\beta) \end{pmatrix}. \tag{5.5.12}$$

Thus we have

$$\text{trace}\,(\mathbf{Q}'\mathbf{Q}^p) = 2\{\cos(2(p+1)\beta) + \sigma \sin(2(p+1)\beta)\}/\sin 2\beta, \tag{5.5.13}$$

where

$$\sigma \equiv -\tfrac{1}{2}(\Delta C/D). \tag{5.5.14}$$

Let us take $(0, \pi/2)$ as the interval of β. The condition (5.5.7) now reduces to

$$|\text{trace}\,(\mathbf{Q}'\mathbf{Q}^p)| \geqq 2, \quad p = 0, 1, 2, \ldots, s - 1. \tag{5.5.15}$$

The necessary and sufficient condition for (5.5.15) is that, for $\sigma < 0$,

$$\sigma < -\cot\{\tau_-(s, \beta)/2\} \sin 2\beta \equiv -\sigma_c^-(s, \beta), \tag{5.5.16}$$

$$\tau_-(s, \beta) \equiv \lim_{\varepsilon \to 0+} \min_{0 \leqq p \leqq s-1} \{2(p+1)\beta - [2(p+1)\beta/\pi - \varepsilon]\,\pi\}, \tag{5.5.17}$$

and for $\sigma > 0$,

$$\sigma > \tan\{\tau_+(s, \beta)/2\} \sin 2\beta \equiv \sigma_c^+(s, \beta), \tag{5.5.18}$$

$$\tau_+(s, \beta) \equiv \max_{0 \leqq p \leqq s-1} \{2(p+1)\beta - [2(p+1)\beta/\pi]\,\pi\}, \tag{5.5.19}$$

where $[x]$ denotes the largest integer not exceeding x.

Using (5.5.14), we find that λ is in a spectral gap of the mixed system if λ lies in the band of the regular system described by **Q** and

$$\Delta C > 2|D|\, \sigma_c^-(s, \beta) \tag{5.5.20}$$

or

$$\Delta C < -2|D|\, \sigma_c^+(s, \beta). \tag{5.5.21}$$

The disordered chain composed of two isotopes considered in § 5.3 is none other than the system of the type we are now considering. In this case it is seen from (1.1.11) that

$$C = 2K, \quad D = -K, \quad E = m^{(0)}\omega^2, \quad \text{or} \quad m^{(1)}\omega^2. \tag{5.5.22}$$

Then

$$\sigma = \tfrac{1}{2}(m^{(0)} - m^{(1)})\, \omega^2/K = -Q^{(1)}m^{(0)}\omega^2/2K. \tag{5.5.23}$$

Further it is readily seen from (3.2.8) that the quantity β is just that defined by (3.2.6), provided that $m^{(0)}$ is taken as the standard mass. Hence we obtain from (5.5.16) and (5.5.18) the condition

$$Q^{(1)} > \sigma_c^-(s, \beta)/2 \sin^2 \beta = \cot \{\tau_-(s, \beta)/2\} \cot \beta \tag{5.5.24}$$

or

$$Q^{(1)} < -\tan \{\tau_+(s, \beta)/2\} \cot \beta. \tag{5.5.25}$$

Thus if the frequency ω is in the interval $(0, 2(K/m^{(0)})^{1/2})$ and (5.5.24) or (5.5.25) is satisfied, it is in one of the gaps of the mixed system. The saw-toothed curve in Fig. 5.12 is just the plot of the function on the right-hand side of (5.5.24).

When β is a rational multiple of $\pi/2$, i.e. if $\beta = (s - t)\, \pi/2s$, where s and t are integers, we have for $s \to \infty$

$$\tau_-(\infty, (s - t)/2s) = \pi/s, \tag{5.5.26}$$

so that (5.5.24) becomes

$$Q^{(1)} > \cot (\pi/2s) \tan (t\pi/2s), \tag{5.5.27}$$

which is identical with the formula (5.3.8) for the critical mass parameter Q_c. Equation (5.5.25) is redundant for $s \to \infty$, since then the right-hand side never becomes larger than -1.

For finite s, there are finite intervals of ω in which the condition (5.5.24) is satisfied, since $\sigma_c^-(s, \beta)$ is continuous in the vicinity of $(s - t)/2s$. The width of such intervals tends, however, to zero as $s \to \infty$, completely in harmony with the observations given in § 5.3.

An entirely similar argument can be given also to the case of disordered Kronig–Penney model, which leads to the formula for the critical potential strength (5.4.8).

Recently Dworin (1965) also attempted to use the theorem of Worpitzky for investigating the spectral gaps in the electronic energy spectra of disordered chains, pointing out thereby that an appropriate choice of the set of transfer matrices would lead to a significant conclusion. It should be remarked, however, that the importance of the freedom of choice of transfer matrices was recognized by Matsuda as early as 1962, and since then it has been fully utilized by Matsuda himself and Hori, as described earlier.

5.6. Peak Structure of the Spectrum of Disordered Systems

The existence of the special points leads one to the belief that the spectrum of the disordered chain should exhibit a remarkably fine structure, having many points at which the spectral density vanishes. The fine structure discovered by Dean *et al.* and Herzenberg and Modinos (§§ 2.4 and 2.5) is thus partly explained. Why there should appear sharp peaks between these special points is, however, still to be elucidated.

As the simplest example, consider a diatomic chain treated in § 5.3. Let us assume that the concentration of light atoms is sufficiently small so that the lattice may be pictured as a "sea" of heavy atoms in which the "islands" composed of one or more light atoms, occasionally with a few heavy atoms inserted between, are floating at sufficiently large mutual distances (Fig. 5.17). To each island corresponds a characteristic transfer matrix. Let $\mathbf{Q}^{(i)}(i = 0, 1, 2, \ldots)$ be the matrix associated with the island of the ith kind, $\mathbf{Q}^{(0)}$ being the matrix of the island of the zeroth kind which consists of a single light atom, and $\mathbf{Q}$ be the matrix of the heavy atom constituting the sea.

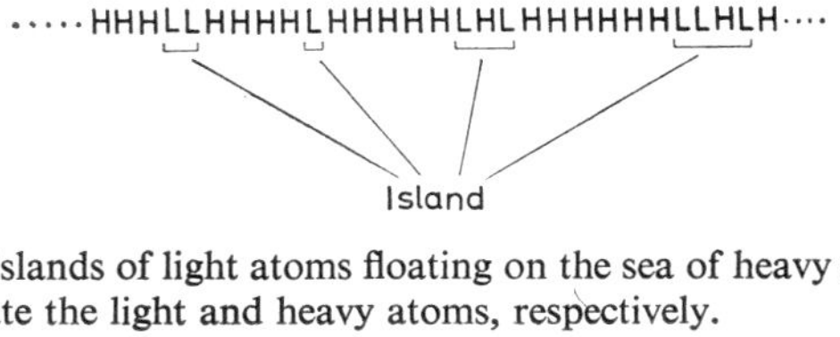

FIG. 5.17. The islands of light atoms floating on the sea of heavy atoms. L and H indicate the light and heavy atoms, respectively.

If an island exists on the sea in isolation, it will give rise to impurity modes with corresponding impurity frequencies. As was explained in Chapter 4, an impurity mode due to the island of the ith kind appears at the β-value for which the sink phase δ_- of $\mathbf{Q}$ is transferred to the source phase δ_+ by the matrix $\mathbf{Q}^{(i)}$. Denote these β-values by $\beta_j^{(i)}$, and assume for simplicity that all the $\beta_j^{(i)}$'s are distinct and sufficiently apart from one another (with respect to both indices).

Since by assumption the islands are sufficiently distant from each other, the phase becomes nearly equal to δ_- just before each island. At an appropriate value of β in a narrow neighbourhood of one of the $\beta_j^{(i)}$'s, e.g. $\beta_n^{(k)}$, a slip occurs at one of the islands of the kth kind, but never occurs at any other island (owing to the assumption of the distinctness of $\beta_j^{(i)}$'s). As the value of β is varied within that neighbourhood, such slips will occur successively at a large number of adjoining β-values, each time at a different island of the kth kind. These values of β give a narrow impurity band characteristic of the island of the kth kind. (For the island of the zeroth kind, this argument was already given in § 4.3.) The same argument, of course, applies also to other β-values $\beta_{m\neq n}^{(k)}$ and to the islands of other kinds. As a result, there appear several narrow impurity bands, each of which corresponds to a particular $\beta_j^{(i)}$.

The peaks appearing in Dean's spectrum should be regarded as the impurity bands produced in the manner explained above. Each peak must therefore correspond closely to an impurity frequency associated with the island of a particular kind. It was already mentioned in § 2.4 that such a correspondence was found by Dean himself empirically.

For the above argument to be valid, it is necessary that the distance between neighbouring islands is larger than the rate of convergence of the phase towards the sink phase δ_-. Taking into account the fact that the width of the peaks in Dean's spectrum is about 0·1 rad, we can estimate from Fig. 4.3 this rate to be about three in the region of medium frequency. This result is in harmony with the finding of Bacon *et al.* (1962) that the impurity frequencies of typical islands are almost unaffected by the presence of more than one island, provided the distance between the neighbouring islands is larger than three.

From Fig. 4.2 it is seen that the rate of convergence is particularly rapid in the region of high frequency and large mass ratio.

Consequently the peak structure must become most conspicuous in such a region. This also is in harmony with the results obtained by Dean (Figs. 2.1–2.6).

Thus we have reached the conclusion that the supposition presented in § 2.4, that even in a disordered chain each island contributes its own impurity frequencies to the spectrum of the whole system almost independently of other islands, in fact corresponds to the truth. It is not strange at all that such an independence of the individual island is preserved up to a fairly high concentration of light atoms.

It is clear that the similar arguments must be valid for any system described by transfer matrices. In any case, the spectrum will exhibit a fine structure if there are distinct islands which lie sufficiently far apart from one another, and at a number of impurity frequencies or energies associated with them, the rate of convergence of the phase on the sea is sufficiently rapid. The fine structure found by Agacy and Borland in the spectrum of a disordered Kronig–Penney model, and by Herzenberg and Modinos in the spectrum of a polymer molecule, can thus be explained in the same way as in the above example. As was mentioned in § 2.5, Agacy and Borland found in fact the correspondence between the peaks and the impurity levels. In the spectrum of Herzenberg and Modinos, the peaks appear only in the region outside the band of *B*-chain, in agreement with the above explanation (Fig. 2.13).

In contrast to these cases, the spectrum of the glass-like lattice calculated by Dean (Figs. 2.9–2.11) has only a smooth structure without any peak. This is natural since in such a case the system cannot be pictured as a sea containing isolated islands.

It would be difficult to deduce such peculiar features of the spectrum of disordered systems as special points and peaks by the perturbational or other approximate methods, which generally predicts only a fairly smooth spectrum having neither narrow peaks nor a number of points of vanishing density (see Chapter 7). At one time this led to a belief that the spectrum should become a smooth one without any singularities once the disorder is introduced in the system. The numerical works of Dean and others and the theoretical argument described in the preceding and the present sections show definitely that some kinds of singularity can persist even when the randomness is introduced into the structure of the system.

5.7. Localization of the Eigenmodes of Disordered Systems

By using the concepts of slip and locking, it can be shown that the amplitude of the eigenmodes of disordered one-dimensional systems must in general be strongly localized. As an example, consider an isotopic chain in which the mass of atoms changes randomly with a continuous distribution function. Such a system is described by a random array of transfer matrices, for example of the form (3.2.8) with continuously distributed γ-values. Let us assume that the distribution of mass has a finite dispersion. Consider a value of β for which only a small fraction of transfer matrices corresponding to the lightest masses are elliptic. According to the result of § 5.1, the sink- and source-phase intervals of the remaining set of hyperbolic matrices must be disjoint.

Owing to the above assumption, the hyperbolic matrices appear in most cases in a long succession, whereas it is a rare event that a long sequence of the elliptic matrices appears. Since the limit-phase intervals are disjoint, the phase must be locked in the sink-phase interval throughout the succession of hyperbolic matrices, possibly excepting the initial few transfers. During the locked state the phase changes, on the average, by just 2π per one transfer. Suppose that such a sequence is interrupted by a short succession of the elliptic matrices. As the average phase change per transfer due to these elliptic matrices is in general only slightly less than 2π, the phase will either continue to remain in the sink-phase interval, or escape from there only temporarily, returning into it as soon as the succession of the elliptic matrices terminates. The average phase change per transfer is thus still locked at 2π.

If, however, an atom with an exceptionally small mass, or an exceptionally long sequence of the elliptic matrices accidentally appears, the phase advance may be sufficiently retarded there so that a slip takes place. When such a slip occurs at a certain atom, the β-value under consideration, β_0, gives an eigenfrequency. On the assumption that there is no degeneracy in eigenfrequencies, the slip can occur only at one place at one time. Therefore beyond this place the phase must remain always in the neighbourhood of the source-phase interval; for the total phase changes at two β-values, of which one is slightly larger and the other slightly smaller than β_0, must differ just by 2π, and moreover, for both β-values the phase must be locked throughout in the neighbourhood of the sink-phase interval, except at the place of the slip, in order not to give rise to an excess slip. Since the total phase change at $\beta = \beta_0$ should have an intermediate value between these amounts of change differing by 2π, the phase must necessarily wander always in the neighbourhood of the source-phase interval beyond the place of the slip. Roberts and Makinson (1962), who presented for the first time an argument of the present type for an electronic system, called this phenomenon *lingering* of the phase.

While the phase is lingering, the state vector becomes, on the average, shorter and shorter. Before the slip takes place, on the contrary, it becomes on the average longer and longer, as the phase remains always in the neighbourhood of the sink-phase interval. Therefore the amplitude of the eigenfunction almost always increases before the slip occurs, and from this place onward it almost always decreases. Thus it is concluded that the eigenfunction is localized around the place at which the slip takes place.

In the above discussion the distribution of the mass was assumed to be continuous. It is clear, however, that this condition is not essential for the above argument. In case the distribution of mass is discrete, and the concept of island is valid, the slip will occur always at one of the islands, and the eigenfunction will be localized around that island. Moreover, the above argument does not depend on any particular model. The localization must occur whenever only a relatively small fraction of the transfer matrices describing the system are elliptic, as far as the disjointness condition is satisfied by the hyperbolic matrices.

Examples of the localization of eigenmodes have already been described in § 2.6. The results obtained by Dean and Bacon (Fig. 2.18) clearly show that the eigenmode is localized just around the island of light atoms at which the slip of the phase takes place, as it should be from the above argument. The localization of the modes of glass-like chains (Fig. 2.20) found by Dean may also be explained by the above mechanism, at least for the modes with sufficiently high frequencies.

The results obtained by Herzenberg and Modinos are also in accordance with the above explanation. As was mentioned in § 2.6, the states which lie outside the band of *B*-chain are localized, whereas those inside that band are not. This is natural, since for the latter states the transfer matrices of both chains are elliptic.

There are, however, some examples in which the eigenmode is localized in spite of the fact that all or a large fraction of the transfer matrices are elliptic at the corresponding energy or frequency. In the isotopically disordered diatomic chain, the calculation of Dean and Bacon shows that the localization begins to occur already at the frequency inside the band of the regular chain of heavy atoms. (It should be noted that for these modes there is no correspondence between the island and the position of localization.) The wave functions in Figs. 2.21 and 2.22, which were calculated by Borland for his liquid-metal model, are well localized. The corresponding energies lie, however, inside the band of the regular chain with the spacing $\langle l_n \rangle$, so that the fractions of elliptic transfer matrices are not necessarily small.

A theory which can explain the localization of eigenmodes without depending upon the above-mentioned conditions was given by Mott and Twose (1961), and later by Borland (1963) in a more refined form. Their theory is, however, confined to some specific electronic systems, and its generalization has not yet been presented.

Recently Taylor (1965) showed that the eigenfunctions cannot be localized even when the system is disordered. His result originates, however, in his too strict definition for the localization, so that the contradiction should be regarded as merely apparent.

CHAPTER 6

Problems on Higher-dimensional Disordered Lattices

6.1. Impurity Frequencies and Peak Structure in Two-dimensional Lattices

It was shown by Hori and Fukushima (1966) that the phase theoretical treatment of the impurity modes described in Chapter 4 can be generalized to the higher-dimensional lattices in a straightforward way. As an example consider a two-dimensional square lattice containing few isotopic impurities. According to (1.4.4), the regular part of the lattice is described by

$$\mathbf{U}_n^{-1} = \mathbf{L} - \mathbf{U}_{n-1}, \tag{6.1.1}$$

where $\mathbf{L}$ is the regular submatrix. The *limit state ratio matrices* are defined to be the matrices which are left invariant by the transformation (6.1.1); i.e. to be the solutions $\mathbf{U}_\pm$ of the equation

$$\mathbf{U}^2 - \mathbf{L}\mathbf{U} + \mathbf{I} = \mathbf{0}. \tag{6.1.2}$$

These are given by

$$\mathbf{U}_\mp = \{\mathbf{L} \pm (\mathbf{L} - 4\mathbf{I})^{1/2}\}/2. \tag{6.1.3}$$

From (1.5.25) the eigenvalues $u_\pm^{(\mu)}$ of $\mathbf{U}_\pm$ are seen to be

$$u_\mp^{(\mu)} = \exp(\pm 2i\beta_\mu), \tag{6.1.4}$$

which are real when the moduli of all the eigenvalues of $\mathbf{L}$ are larger than 2. From (1.5.25) and (1.5.24), this is seen to correspond to the case in which ω is larger than the maximum frequency $4(K + K')/m$ of the regular lattice described by $\mathbf{L}$, and we have to put $\beta_\mu = (\pi/2) + i\varepsilon_\mu$ (ε_μ: real). Then the eigenvalues $u_\pm^{(\mu)}$ and χ_μ become

$$u_\mp^{(\mu)} = -\exp(\mp 2\varepsilon_\mu) \tag{6.1.5}$$

and

$$\chi_\mu = -2\cosh 2\varepsilon_\mu \tag{6.1.6}$$

respectively. From (6.1.6) we see that in the present case $\mathbf{L}$ is always smaller than $-2\mathbf{I}$.

In order to justify the order of double sign adopted in (6.1.4) and (6.1.5), we should go back to (1.3.4) with the regular transfer matrix (1.6.1):

$$\begin{pmatrix} \mathbf{V}_n \\ \mathbf{V}_{n+1} \end{pmatrix} = \begin{pmatrix} \mathbf{0} & \mathbf{I} \\ -\mathbf{I} & \mathbf{L} \end{pmatrix} \begin{pmatrix} \mathbf{V}_{n-1} \\ \mathbf{V}_n \end{pmatrix}. \tag{6.1.7}$$

The limit vectors $(V_{\pm}^{(1)}, V_{\pm}^{(2)})^T$ are the eigenvectors of the regular transfer matrix corresponding to the eigenvalues (1.6.2), where we now have

$$|\boldsymbol{\theta}_-| > |\boldsymbol{\theta}_+|. \tag{6.1.8}$$

Since

$$\mathbf{U}_{\pm} \equiv \mathbf{V}_{\pm}^{(1)}\mathbf{V}_{\pm}^{(2)-1} = \boldsymbol{\theta}_{\pm}^{-1} = \tfrac{1}{2}\{\mathbf{L} \mp (\mathbf{L}^2 - 4\mathbf{I})^{1/2}\}, \tag{6.1.9}$$

the order of the double sign in (6.1.3), (6.1.4), and (6.1.5) is seen to be right.

Now suppose that there is an isotopic impurity with the mass m' at $n = i$, $m = j$. If i is sufficiently large, $\mathbf{U}_i$ must be approximately equal to $\mathbf{U}_-$, and consequently $\mathbf{U}_{i+1}$ must be given by

$$\mathbf{U}_{i+1}^{-1} = \mathbf{L}^{(j)} - \mathbf{U}_-, \tag{6.1.10}$$

where $\mathbf{L}^{(j)}$ is the *impurity submatrix* given by

$$\mathbf{L}^{(j)} \equiv \mathbf{L} - \omega^2 \Delta\mathbf{M}^{(j)},$$

$$\Delta\mathbf{M}^{(j)} \equiv \begin{pmatrix} 0 & & & & \\ & \ddots & & & \\ & & 0 & & \\ & & & (m'-m)/K & & \\ & & & & 0 & \\ & & & & & \ddots \\ & & & & & & 0 \end{pmatrix} \text{... the } j\text{th row.} \tag{6.1.11}$$

On the other hand, if we transfer backward from the opposite end of the lattice ($n = -\infty$), $\mathbf{U}_{i+1}$ must become nearly equal to $\mathbf{U}_+$, since $\mathbf{U}_+$ plays the role of the sink state ratio matrix in the backward transfer. This does not mean, however, that the matrix equation $\mathbf{L}^{(j)} - \mathbf{U}_- = \mathbf{U}_-$ must be valid, but only means that the matrices $\mathbf{U}_+$ and $(\mathbf{L}^{(j)} - \mathbf{U}_-)^{-1}$ should give the same displacement vector $\mathbf{v}_{i+1}$ at $n = i + 1$, that is, there should exist a vector $\mathbf{z}$ such that

$$\mathbf{U}_-\mathbf{z} = (\mathbf{L}^{(j)} - \mathbf{U}_-)\mathbf{z}. \tag{6.1.12}$$

The initial vector corresponding to this displacement vector is $\mathbf{V}_i^{-1}\mathbf{z}$, since $\mathbf{U}_{i+1}^{-1}\mathbf{z} = \mathbf{V}_{i+1}\mathbf{V}_i^{-1}\mathbf{z}$. Thus we get the equation

$$|2\mathbf{U}_- - \mathbf{L}^{(j)}| = 0, \tag{6.1.13}$$

as the condition for an impurity frequency to appear.

In case the edges $m = 1$ and $m = M$ are fixed, we readily obtain, by using (1.5.18), the matrix $\Delta\mathbf{M}^{(j)}$ in the representation in which $\mathbf{L}$ is diagonalized:

$$\frac{2Qm}{K(M+1)}\begin{pmatrix} \sin 2\theta_j \sin 2\theta_j & \sin 4\theta_j \sin 2\theta_j & \ldots \sin 2M\theta_j \sin 2\theta_j \\ \sin 2\theta_j \sin 4\theta_j & \sin 4\theta_j \sin 4\theta_j & \ldots \sin 2M\theta_j \sin 4\theta_j \\ \cdots\cdots\cdots & & \\ \cdots\cdots\cdots & & \\ \sin 2\theta_j \sin 2M\theta_j & \sin 4\theta_j \sin 2M\theta_j & \ldots \sin 2M\theta_j \sin 2M\theta_j \end{pmatrix}, \tag{6.1.14}$$

where we have defined the mass parameter Q by $Q \equiv (m'/m) - 1$, taking into account that we are now using the host atom as the basic one (the wave-number parameters β have been defined by (1.5.24)). Then from (6.1.5) and (6.1.6), (6.1.13) becomes

$$\begin{vmatrix} \sin^2 2\theta_j + \dfrac{(M+1)K}{\omega^2 mQ}\sinh 2\varepsilon_1 & \sin 4\theta_j \sin 2\theta_j & \ldots \sin 2M\theta_j \sin 2\theta_j \\ \sin 2\theta_j \sin 4\theta_j & \sin^2 4\theta_j + \dfrac{(M+1)K}{\omega^2 mQ}\sinh 2\varepsilon_2 & \ldots \sin 2M\theta_j \sin 4\theta_j \\ \cdots & \cdots & \cdots \\ \sin 2\theta_j \sin 2M\theta_j & & \sin^2 2M\theta_j + \dfrac{(M+1)K}{\omega^2 mQ}\sinh 2\varepsilon_M \end{vmatrix} = 0. \tag{6.1.15}$$

Writing out the determinant, all the terms vanish except those containing the factor $\sinh 2\varepsilon_\mu$ M or $M - 1$ times, so that we have

$$1 + \sum_{\mu=1}^{M} \frac{\omega^2 mQ}{K(M+1)} \frac{1}{\sinh 2\varepsilon_\mu}\left(\sin \frac{\mu j\pi}{M+1}\right)^2 = 0. \tag{6.1.16}$$

In the one-dimensional case, where $M = 1$ and $K' = 0$, this reduces to

$$1 + Q \coth \varepsilon = 0, \tag{6.1.17}$$

which coincides with (4.3.18), as can easily be verified.

The displacement vector at $n = i$ is given by

$$\mathbf{V}_i \mathbf{V}_i^{-1} \mathbf{z} = \mathbf{z}, \tag{6.1.18}$$

which is the eigenvector of the matrix $2\mathbf{U}_- - \mathbf{L}^{(j)}$ corresponding to the eigenvalue zero. The components of $\mathbf{z}$ are easily obtained as

$$z_m = \sin 2\theta_j \sin 2m\theta_j / \sinh 2\varepsilon_m . \tag{6.1.19}$$

Returning to the original representation, (6.1.19) becomes

$$\begin{aligned} z_m &= \sum_k \sin 2m\theta_k \sin 2\theta_j \sin 2k\theta_j / \sinh 2\varepsilon_k \\ &= \sin \frac{j\pi}{M+1} \sum_k \sin \frac{kj\pi}{M+1} \sin \frac{mk\pi}{M+1} / \sinh 2\varepsilon_k , \end{aligned} \tag{6.1.20}$$

from which it is seen that the displacement is maximum at $m = j$ and decays rapidly towards both sides. Since $\mathbf{U}_n^{-1}\mathbf{v}_{n-1} = \mathbf{V}_n \mathbf{V}_{n-1}^{-1} \mathbf{v}_{n-1} = \mathbf{V}_n \mathbf{v}_1 = \mathbf{v}_n$, $\mathbf{U}_n^{-1}$ plays the role of the transfer matrix for the displacement vector. From (6.1.5) we see that for $n \leqq i$, the displacement increases with n, whereas for $n > i$ it decreases with n. Thus the normal mode associated with the frequency given by (6.1.16) is completely localized around the impurity atom.

When there are two impurities with mass m' at $m = k$, $m = j$ on the line $n = i$, we obtain in the same way the equation

$$\begin{vmatrix} \begin{array}{l} \sin^2 2\theta_k + \sin^2 2\theta_j \\ + \dfrac{(M+1)K}{\omega^2 mQ} \sinh 2\varepsilon_1 \end{array} & \begin{array}{l} \sin 4\theta_k \sin 2\theta_k \\ + \sin 4\theta_j \sin 2\theta_j \end{array} & \begin{array}{l} \ldots \sin 2M\theta_k \sin 2\theta_k \\ + \sin 2M\theta_j \sin 2\theta_j \end{array} \\[2ex] \begin{array}{l} \sin 2\theta_k \sin 4\theta_k \\ + \sin 2\theta_j \sin 4\theta_j \end{array} & \begin{array}{l} \sin^2 4\theta_k + \sin_2 4\theta_j \\ + \dfrac{(M+1)K}{\omega^2 mQ} \sinh 2\varepsilon_2 \end{array} & \begin{array}{l} \ldots \sin 2M\theta_k \sin 4\theta_k \\ + \sin 2M\theta_j \sin 4\theta_j \end{array} \\[2ex] \begin{array}{l} \sin 2\theta_k \sin 2M\theta_k \\ + \sin 2\theta_j \sin 2M\theta_j \end{array} & \ldots & \begin{array}{l} \sin^2 2M\theta_k + \sin^2 2M\theta_j \\ + \dfrac{(M+1)K}{\omega^2 mQ} \sinh 2\varepsilon_M \end{array} \end{vmatrix} = 0, \tag{6.1.21}$$

instead of (6.1.15). This calculated out to be

$$\left\{1 + \sum_{\mu=1}^{M} \frac{\omega^2 mQ}{(M+1)K} \frac{1}{\sinh 2\varepsilon_\mu} \left(\sin \frac{\mu k\pi}{M+1}\right)^2\right\}$$

$$\times \left\{1 + \sum_{\nu=1}^{M} \frac{\omega^2 mQ}{(M+1)K} \frac{1}{\sinh 2\varepsilon_\nu} \left(\sin \frac{\nu j\pi}{M+1}\right)^2\right\} - \frac{\omega^4 m^2 Q^2}{(M+1)^2 K^2}$$

$$\times \sum_{\mu,\nu}^{M} \frac{1}{\sinh 2\varepsilon_\mu \sinh 2\varepsilon_\nu} \sin \frac{\mu k\pi}{M+1} \sin \frac{\nu k\pi}{M+1} \sin \frac{\mu j\pi}{M+1}$$

$$\times \sin \frac{\nu j\pi}{M+1} = 0. \tag{6.1.22}$$

For sufficiently large $|k - j|$, the second term of this equation becomes very small, so that the equation is approximately factorized into two identical equations, which furthermore coincide with (6.1.16). This means that the impurity frequencies rapidly draw together and approach the impurity frequency associated with a single impurity, as the separation between the impurities becomes large.

Next consider the case in which two impurities are at $n = i$, $m = k$ and $n = i + 1$, $m = j$. The matrix $\mathbf{U}_{i+1}^{-1}$ is, as before, given by

$$\mathbf{U}_{i+1}^{-1} = \mathbf{L}^{(k)} - \mathbf{U}_{-}. \tag{6.1.23}$$

If we transfer backward from the opposite end of the lattice, the matrix at $n = i + 1$ becomes

$$\mathbf{U}_{i+1}^{-1\prime} = (\mathbf{L}^{(j)} - \mathbf{U}_{i+2}^{-1})^{-1} = (\mathbf{L}^{(j)} - \mathbf{U}_{-})^{-1}, \tag{6.1.24}$$

since $\mathbf{U}_n = \mathbf{L}_n - \mathbf{U}_{n+1}^{-1}$. The condition for $\mathbf{U}_{i+1}^{-1}$ and $\mathbf{U}_{i+1}^{-1\prime}$ to give the same displacement vector $\mathbf{v}_{i+1}$ is that there exists a vector $\mathbf{z}$ such that

$$(\mathbf{U}_{i+1}^{-1} - \mathbf{U}_{i+1}^{-1\prime})\,\mathbf{z} = \mathbf{0}, \quad \text{or} \quad |\mathbf{U}_{i+1}^{-1} - \mathbf{U}_{i+1}^{-1\prime}| = 0. \tag{6.1.25}$$

For the matrix $\mathbf{U}_{i+1}^{-1} - \mathbf{U}_{i+1}^{-1\prime}$, we find

$$\begin{aligned}\mathbf{U}_{i+1}^{-1} - \mathbf{U}_{i+1}^{-1\prime} &= \mathbf{L}^{(k)} - \mathbf{U}_{-} - (\mathbf{L}^{(j)} - \mathbf{U}_{-})^{-1} \\ &= (\mathbf{L}^{(j)} - \mathbf{U}_{-})^{-1}\{\mathbf{L}^{(j)} - 2\mathbf{U}_{-} - \omega^2 \mathbf{U}_{+} \Delta\mathbf{M}^{(k)} \mathbf{U}_{-}\}\,\mathbf{U}_{+}.\end{aligned} \tag{6.1.26}$$

By a calculation entirely similar to the above, (6.1.25) turns out to be

$$\left\{1+\sum_{\mu=1}^{M}\frac{\omega^2 mQ}{(M+1)K}\frac{1}{\sinh 2\varepsilon_\mu}\left(\sin\frac{\mu k\pi}{M+1}\right)^2\right\}\left\{1+\sum_{\nu=1}^{M}\frac{\omega^2 mQ}{(M+1)K}\right.$$
$$\left.\times\frac{1}{\sinh 2\varepsilon_\nu}\left(\sin\frac{\nu j\pi}{M+1}\right)^2\right\}-\frac{\omega^4 m^2 Q^2}{(M+1)^2K^2}\sum_{\mu,\nu}^{M}\frac{\exp 2(\varepsilon_\nu-\varepsilon_\mu)}{\sinh 2\varepsilon_\mu \sinh 2\varepsilon_\nu}$$
$$\times \sin\frac{\mu k\pi}{M+1}\sin\frac{\nu k\pi}{M+1}\sin\frac{\mu j\pi}{M+1}\sin\frac{\nu j\pi}{M+1}=0. \tag{6.1.27}$$

This equation has completely the same structure as (6.1.22), and shows that for sufficiently large $|k-j|$, the impurity frequencies are almost equal to that due to one impurity.

Similar argument can be repeated for any number of impurities, and one is led to the conclusion that, in any case, there appears a narrow impurity band centred at the impurity frequency associated with a single impurity, provided that the impurities are sufficiently far apart from one another.

Returning to the case of one impurity, let us next suppose that the impurity is on the free edge. Then $i = N$, and according to (1.5.5), there must exist a vector such that

$$\mathbf{U}_{N+1}^{-1}\mathbf{z} = \mathbf{z}. \tag{6.1.28}$$

Therefore the equation for the impurity frequency is

$$|\mathbf{L}^{(j)} - \mathbf{U}_- - \mathbf{I}| = 0. \tag{6.1.29}$$

By a calculation similar to the previous one, this turns out to be

$$1+\frac{\omega^2 mQ}{2K}\left\{1-\frac{2}{M+1}\sum_{\mu=1}^{M}\tan\varepsilon_\mu\left(\sin\frac{\mu j\pi}{M+1}\right)^2\right\}=0. \tag{6.1.30}$$

In the one-dimensional case, this reduces to

$$1+2Q\cosh\varepsilon\,(\cosh\varepsilon-\sinh\varepsilon)=0. \tag{6.1.31}$$

From this equation it is seen that if the impurity is on a free end, m' must be less than $m/2$ for an impurity frequency to appear. Thus the condition for the appearance of the impurity frequency becomes severer than in the previous case, in which it was sufficient that m' is less than m.

We have assumed above that the boundaries $m = 1$ and $m = M$ are fixed. The calculation runs in an entirely similar way when these are free boundaries. One has only to substitute the eigenvalues χ_μ and eigenvectors $Z_m^{(\mu)}$ given by (1.5.20) and (1.5.21), respectively, for those given by (1.5.17) and (1.5.18). In the case of one impurity with mass m' at $n = i$, $m = j$, the eigenfrequency equation for the impurity mode becomes now

$$1 + \frac{\omega^2 mQ}{2MK} \sum_{\mu=0}^{2M-1} \frac{1}{\sinh 2\varepsilon_\mu} \cos^2 \left\{ \frac{(2j-1)\,\mu\pi}{2M} \right\} = 0, \tag{6.1.32}$$

where $\varepsilon_{2M-m} \equiv \varepsilon_m$. In the one-dimensional case, this reduces to (6.1.17), as it should be. When the impurity lies on the free edge $i = N$, we have

$$1 + \frac{\omega^2 mQ}{2K} \left[1 - \frac{1}{M} \sum_{\mu=0}^{2M-1} \tanh \varepsilon_\mu \cos^2 \left\{ \frac{(2j-1)\,\mu\pi}{2M} \right\} \right] = 0. \tag{6.1.33}$$

instead of (6.1.32). For the one-dimensional case, this reduces naturally to (6.1.31).

The corresponding formulae can, of course, be derived also for the case of cyclic boundary condition by simply using the eigenvalues and eigenvectors given by (1.5.22) and (1.5.23) respectively, in place of the above ones.

As before, the normal mode associated with the frequency given by (6.1.32) or (6.1.33) comes out to be localized around the impurity, and it turns out that, in the case of several impurities, the impurity frequencies are nearly equal to that associated with a single impurity, provided that they are sufficiently apart from one another. This gives the explanation of the fact found by Dean and Bacon and mentioned in § 2.4 that also the spectrum of the two-dimensional system has a peak structure and, moreover, each peak corresponds well to an impurity frequency associated with a particular kind of island.

6.2. Line Impurities

The phase theory can be applied also to the case in which a whole line $n = i$, or a whole edge $n = N$ is occupied by impurity atoms of the same kind (*line impurities*). Suppose that the line $n = i$ is composed exclusively of the isotope with mass m'. In this case the matrix $\mathbf{L}^{(j)}$ appearing in the preceding section should be replaced by

$$\mathbf{L}' \equiv \mathbf{L} - \omega^2 mQ\mathbf{I}/K. \tag{6.2.1}$$

The matrix $\mathbf{L}'_n$ commutes with $\mathbf{L}$, so that its eigenvalues are given by

$$2\cos 2\beta_\mu - \omega^2 mQ/K. \tag{6.2.2}$$

Contrary to the case considered in § 6.1, now it is not necessary that all the β_μ's are complex, for all the matrices appearing in the problem commute with one another, so that the problem essentially becomes one-dimensional; we have only to consider each time one particular eigenvalue with the corresponding normal mode, independently of other eigenvalues and the associated normal modes. It is sufficient that this particular mode corresponds to a complex β_μ; in other words, it is sufficient that the submatrix corresponding to the normal mode under consideration is hyperbolic.

Equation (6.1.13) now turns out to be

$$2\exp(2i\beta_\mu) - 2\cos 2\beta_\mu + \omega^2 mQ/K = 0. \tag{6.2.3}$$

Putting $\beta_\mu = (\pi/2) + i\varepsilon_\mu$, this becomes

$$1 + \omega^2 mQ/2K \sinh 2\varepsilon_\mu = 0, \tag{6.2.4}$$

which is to give the impurity frequency. In the one-dimensional case, this is reduced to

$$1 + Q \coth \varepsilon = 0, \tag{6.2.5}$$

which coincides with (6.1.17), as it should.

The displacement vector at $n = i$ is as before given by the eigenvector of the matrix $2\mathbf{U}_- - \mathbf{L}'$. It is, however, evidently the same as that of $\mathbf{L}$, given by (1.5.18) or (1.5.21) or (1.5.23), according as the boundary condition on $m = 1$ and $m = M$ is fixed or free or periodic. In any case it is not localized but sinusoidal. By the argument which is completely the same as that given below the expression (6.1.20), however, it is easily seen that the displacement increases exponentially with n for $n \leqq i$, whereas it decreases exponentially with n for $n > i$. Thus the normal modes associated with the eigenfrequencies obtained from (6.2.4) are strongly localized in the x-direction, but not localized in the y-direction.

The eigenfrequencies of these modes are not necessarily larger than $\omega_{\max} = 4(K + K')/m$. For in the present case some of β_μ's may be real, so that in the case of free boundary condition, for example, the frequency is given by

$$\omega^2 = 4K \sin^2 \beta_\mu/m + 4K' \sin^2 (\mu\pi/2M)/m \qquad (6.2.6)$$

for each value of μ corresponding to a real β_μ.

Next consider the case in which the free edge $n = N$ is composed of impurities (*impurity edge*). Equation (6.1.29) here becomes

$$2\cos 2\beta_\mu - \omega^2 mQ/K - \exp(2i\beta_\mu) - 1 = 0, \qquad (6.2.7)$$

which reduces to

$$1 + \omega^2 mQ/2K \exp \varepsilon_\mu \cosh \varepsilon_\mu = 0. \qquad (6.2.8)$$

In the one-dimensional case, this comes out to be the same as (6.1.31), as it should be.

Also in the present case it can easily be seen that the normal modes corresponding to the roots of (6.2.8) are localized in the x-direction; the decay exponentially as the distance from the edge $n = N$ increases. Such modes are called *surface modes*. The critical values Q_μ for the appearance of the complex roots are obtained by putting $\varepsilon_\mu = 0$ in (6.2.8), that is,

$$Q_\mu = -\tfrac{1}{2}K/\{K + K' \sin^2 (\mu\pi/2M)\}. \qquad (6.2.9)$$

From this formula it follows that

$$-\tfrac{1}{2} = Q_0 < Q_1 < \dots < Q_{M-1} < -\tfrac{1}{2}(1 - p^2), \qquad (6.2.10)$$

where

$$p \equiv \{K'/(K + K')\}^{1/2}. \qquad (6.2.11)$$

This means that the frequencies corresponding to the surface modes appear more easily from the high-frequency region than from the low-frequency part of the band of the regular lattice.

In the limit $M \to \infty$, the frequencies of surface modes form a continuous band, which may overlap with the main band which is composed of the frequencies of the modes not localized in any direction (*extended modes*) and occupy the same frequency region as the band of the regular lattice. Fukushima (1965) calculated from (6.2.8) the spectral density in the band of the surface-mode frequencies. It was found to be given by

$$D(\omega^2) = \frac{m}{2\pi K'}\left\{1 + Q - \frac{Q}{1 + (\omega^2 mQ)/K}\right\}$$
$$\times\left[1 - \frac{M\omega^2}{2K'}\left\{1 + Q - \frac{Q}{1 + (\omega^2 mQ)/K}\right\}^2\right]^{-1/2}. \quad (6.2.12)$$

This function is plotted in Fig. 6.1 for several values of γ and K.

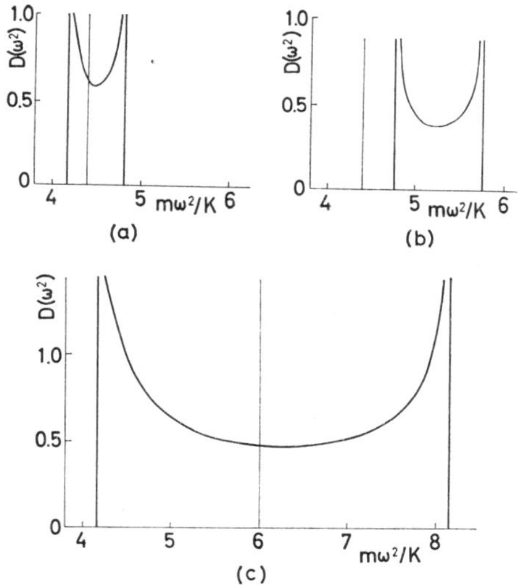

FIG. 6.1. The spectral density in the band of surface-mode frequencies for the two-dimensional lattice with an impurity edge. (a) $\gamma \equiv K'/K = 0{\cdot}1$, $Q = -0{\cdot}6$; (b) $\gamma = 0{\cdot}1$, $Q = -0{\cdot}7$; (c) $\gamma = 0{\cdot}5$, $Q = -0{\cdot}6$. The top of the main band is indicated by a vertical line. In the cases (a) and (c), the band of surface-mode frequencies overlaps with the main band. The shape of the density curve is very similar to that of the one-dimensional regular chain. (After Fukushima (1965).)

Completely similar arguments can be repeated for the three-dimensional case as well. In general, when the impurities occupy a whole subspace with one- or two-dimensional extensions (*extended impurities*), there appear normal modes which are localized in the direction normal to that subspace, but not localized (extended) in the subspace itself. The frequencies of these impurity modes may lie inside the band of the regular lattice, forming the bands which may overlap with the main band. The same conclusion was obtained in a different way by Lengeler and Ludwig (1964) and Kobori (1965) (see the Bibliographical Notes).

6.3. Impurities on a Surface

We have already considered in § 6.1 the impurity mode associated with an impurity on the edge $n = N$ of the square lattice. Fukushima (1965) investigated a somewhat more complex case in which an atom of mass m' is *adsorbed* on the free edge $n = N$ at the site $m = j$, interacting with the surface atom through the force constant K'' (Fig. 6.2). Denote the displacement of the adsorbed atom by v. Then the equation of motion for v_{Nm} and v are

$$-m\omega^2 v_{Nm} = K''(v - v_{Nm}) - K(v_{Nm} - v_{N-1,m}) + K'(v_{N,m+1} - 2v_{Nm} + v_{N,m-1}) \qquad (6.3.1\,\mathrm{a})$$

and

$$-m'\omega^2 v = K''(v_{Nm} - v) \qquad (6.3.1\,\mathrm{b})$$

respectively. The equation of motion for other atoms are not affected by the presence of the adsorbed atom.

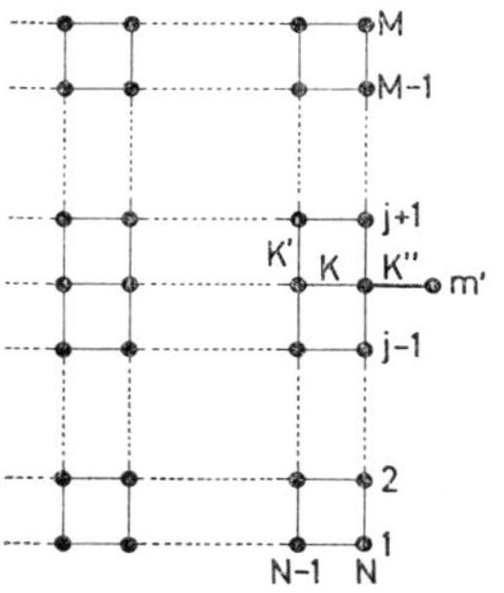

FIG. 6.2. The two-dimensional lattice in which an extra atom with mass m' is adsorbed on its edge $n = N$.

Eliminating v from these equations, we get

$$m(1 + Q')\,\omega^2 v_{Nm} = K(v_{N-1,m} - v_{Nm}) + K'(v_{N,m+1} - 2v_{Nm} + v_{N,m-1}), \qquad (6.3.2)$$

where

$$Q' \equiv K(1 + q)(1 + Q)/\{K(1 + q) - m\omega^2(1 + Q)\}, \qquad (6.3.3)$$

and $q \equiv (K''/K) - 1$. Thus our system is equivalent to that in which the surface atom at (N, m) is replaced by an isotopic impurity with mass $m'' = m(1 + Q')$. Therefore, the equation for the impurity frequency must be the same as (6.1.33) except that Q is replaced by Q':

$$1 + \frac{\omega^2 m Q'}{2K}\left[1 - \frac{1}{M}\sum_{\mu=0}^{2M-1}{}' \tanh \varepsilon_\mu \cos^2\left\{\frac{(2j-1)\,\mu\pi}{2M}\right\}\right] = 0. \quad (6.3.4)$$

Starting from this equation, Fukushima examined the behaviour of the critical value Q_c for the appearance of an impurity mode. Using (1.5.24), (6.3.4) can be written

$$1 + \frac{\omega^2 m Q'}{2K}\left[1 - \frac{1}{M}\sum_{\mu=0}^{2M-1}{}' \left\{1 - \frac{4K}{m\omega^2 - 4K'\sin^2(\mu\pi/2M)}\right\}^{1/2}\right.$$
$$\left.\times \cos^2\left\{\frac{(2j-1)\,\mu\pi}{2M}\right\}\right] = 0. \quad (6.3.5)$$

If we put $m\omega^2 = 4(K + K')$ in (6.3.5), it becomes the equation which determines Q_c:

$$1 + \frac{2(1+q)(1+Q_c)}{(1+q)(1-p^2) - 4(1+Q_c)}\left[1 - \frac{1}{M}\right.$$
$$\left.\times \sum_{\mu=0}^{2M-1}{}' \frac{p|\cos(\mu\pi/2M)|}{\{1 - p^2\sin^2(\mu\pi/2M)\}^{1/2}} \cos^2\left\{\frac{(2j-1)\,\mu\pi}{2M}\right\}\right] = 0, \quad (6.3.6)$$

where $p^2 \equiv K'/(K + K') < 1$. For $p = q = 0$, (6.3.6) is simplified to

$$Q_c = -\tfrac{1}{2}. \quad (6.3.7)$$

Since $p = 0$ corresponds to the case of one-dimensional lattice, this is in harmony with the conclusion given below (6.1.31).

For other values of p, the value of Q_c depends on the position of the impurity j. Let us examine four special cases:

(1) An isotopic impurity is adsorbed at the centre of the edge, i.e. $2j - 1 = M$ and $K'' = K$.

In this case (6.3.6) becomes

$$1 + \frac{2(1+Q_c)}{1 - p^2 - 4(1+Q_c)}\left[1 - \frac{1}{M}\sum_{\mu=0}^{2M-1}{}' \frac{p}{\{1 - p^2\sin^2(\mu\pi/2M)\}^{1/2}}\right.$$
$$\left.\times \left|\cos\frac{\mu\pi}{2M}\right|\right] = 0, \quad (6.3.8)$$

where the prime indicates the summation over even μ only. When M is large, the sum may be approximated by an integral, so that

$$1 + \frac{2(1+Q_c)}{1-p^2-4(1+Q_c)}\left[1 - \frac{\pi}{2}\int_0^{\pi/2} \frac{d\varphi\, p\cos\varphi}{\{1-p^2\sin^2\varphi\}^{1/2}}\right] = 0, \tag{6.3.9}$$

from which we obtain

$$Q_c = \tfrac{1}{2}(1-p^2)\,\{1 + (\pi/2)\sin^{-1}p\}^{-1} - 1. \tag{6.3.10}$$

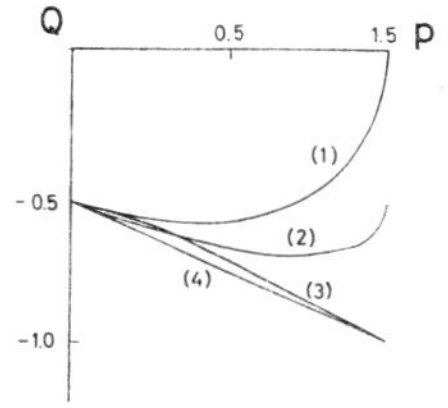

FIG. 6.3. The critical value Q_c for the appearance of the impurity mode as a function of p, for the cases in which (1) an isotopic impurity is on the edge $n = N$ at $m = (M+1)/2$, (2) an isotopic impurity is on the edge at $m = 1$, (3) an isotopic extra atom is adsorbed at $m = (M+1)/2$, and (4) an extra atom is adsorbed at $m = 1$. (After Fukushima (1965) and Asahi (1964).)

The variation of Q_c as a function of p is shown in Fig. 6.3 (curve 3).

(2) An isotopic impurity is adsorbed at the corner, i.e. $j = 1$, $K'' = K$. By a calculation similar to the above, we get

$$Q_c = \frac{\tfrac{1}{2}(1-p^2)\,(1+(\pi/2))}{(1-p^2)^{1/2}/p - \{(1-2p^2)\sin^{-1}p\}/p^2} - 1, \tag{6.3.11}$$

which is plotted also in Fig. 6.3 (curve 4). From the curves 3 and 4 it is seen that the impurity mode appears more easily in the case of adsorption at the centre than in the case of adsorption at the corner.

(3) A non-isotopic impurity is adsorbed at the centre of the edge, i.e. $2j - 1 = M$ and $K'' \neq K$. In this case we have

$$Q_c = \frac{\tfrac{1}{2}(1+q)\,(1-p^2)}{2-(1+q)\,(1-(2/\pi)\sin^{-1}p)} - 1. \tag{6.3.12}$$

(4) A non-isotopic impurity is adsorbed at the corner, i.e. $j = 1$ and $K'' \neq K$. In this case it follows

$$Q_c = -1 + \frac{\frac{1}{2}(1+q)(1-p^2)}{2-(1+q)\,[1-(2/\pi)\{(1-p^2)^{1/2}/p-(1-2p^2)\sin^{-1}p/p^2\}]}. \tag{6.3.13}$$

The variation of Q_c as a function of p is shown in Fig. 6.4 for the latter two cases. The curves become flat as q tends to -1, showing that when the adsorbed atom is weakly coupled with the lattice,

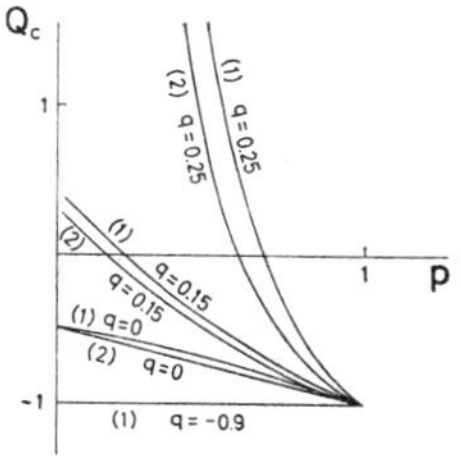

FIG. 6.4. The critical value Q_c for the appearance of the impurity mode as a function of p for the cases in which a non-isotopic atom is adsorbed (1) at $m = (M+1)/2$ and (2) at $m = 1$, for several values of q. (After Fukushima (1965).)

the impurity mode appears only for a very light adatom. It is seen that also in the non-isotopic case the impurity mode appears more easily in the case of adsorption at the centre than in the case of adsorption at the corner.

In a similar manner Asahi (1964) and Asahi and Hori (1963) examined the behaviour of Q_c for the case of isotopic impurity *on* the edge. Starting from (6.1.33), they obtained the formula

$$Q_c = -\tfrac{1}{2}(1-p^2)/\{1-(2/\pi)\sin^{-1}p\}, \tag{6.3.14}$$

for the case $2j-1 = M$, and

$$Q_c = \frac{-\frac{1}{2}(1-p^2)}{1-(2/\pi)\{(1-p^2)^{1/2}p-(1-2p^2)\,|p^2\sin^{-1}p|\}}, \tag{6.3.15}$$

for the case $j = 1$. These functions are plotted in Fig. 6.3 (curves 1 and 2 respectively). Also here the impurity mode appears more easily in the case in which the impurity lies at the centre of the edge than in the case in which it is at the corner.

When $K' < K$, the value of p^2 is smaller than $\frac{1}{2}$, approaching zero as $K' \to 0$. This limit corresponds to the case in which the lattice is decomposed into independent linear chains extended in the x-direction. Hence it is natural that here both curves 1 and 2 give the same value $Q_c = -\frac{1}{2}$, which agrees with the result obtained above in the one-dimensional case. On the other hand, p^2 is larger than $\frac{1}{2}$ when $K' > K$, approaching unity as $K \to 0$. This limit corresponds to the case in which the lattice is decomposed into linear chains extended in the y-direction. Therefore it is a matter of course that for $p^2 = 1$ the curve 2 gives $Q_c = -\frac{1}{2}$, whereas the curve 1 gives $Q_c = 0$. That the curve 1 gives higher values of Q_c in the right half than in the left half of the figure means that, if the force constant in the longitudinal direction is larger than that in the transverse direction, as is usually the case, the vibration in the direction of the edge is more easily localized than that in the direction normal to the edge for the impurity situated at its centre.

6.4. Method of Transfer Matrix

We have seen thus far that the phase theory enables us to obtain the equation for the frequencies of impurity modes in a very simple manner. It fails, however, for the investigation of the extended mode, since for such a mode the transfer matrix as a whole or the corresponding submatrix is elliptic, so that no limit phase matrix or submatrix exists. We are obliged, then, to return to the description by the transfer matrix itself.

As an example of the use of the transfer matrix, consider a square lattice with free boundaries, in which the whole edge $n = N$ is composed of the isotopes with mass m'. Then the state vectors at $n = 1$ and $n = N + 1$ are connected by the product $\mathbf{T}'\mathbf{T}^{N-1}$:

$$\mathbf{X}_{N+1} = \mathbf{T}'\mathbf{T}^{N-1}\mathbf{X}_1, \tag{6.4.1}$$

where $\mathbf{X}_n \equiv (\mathbf{V}_{n-1}, \mathbf{V}_n)^T$, $\mathbf{T}$ is the matrix given by (1.6.1), and

$$\mathbf{T}' \equiv \begin{pmatrix} \mathbf{0} & \mathbf{I} \\ -\mathbf{I} & \mathbf{L}' \end{pmatrix}. \tag{6.4.2}$$

In the representation which diagonalizes **T**, (6.4.1) becomes

$$\begin{pmatrix} \boldsymbol{\theta}_-\mathbf{V}_N - \mathbf{V}_{N+1} \\ \\ -\boldsymbol{\theta}_+\mathbf{V}_N + \mathbf{V}_{N+1} \end{pmatrix} = \begin{pmatrix} \boldsymbol{\theta}_+^N\left\{1 - \dfrac{\Delta\mathbf{L}}{(\boldsymbol{\theta}_- - \boldsymbol{\theta}_+)}\right\} & -\dfrac{\Delta\mathbf{L}\boldsymbol{\theta}_-^N}{(\boldsymbol{\theta}_- - \boldsymbol{\theta}_+)} \\ \dfrac{\Delta\mathbf{L}\boldsymbol{\theta}_+^N}{(\boldsymbol{\theta}_- - \boldsymbol{\theta}_+)} & \boldsymbol{\theta}_-^N\left\{1 + \dfrac{\Delta\mathbf{L}}{(\boldsymbol{\theta}_- - \boldsymbol{\theta}_+)}\right\} \end{pmatrix}$$

$$\times \begin{pmatrix} \boldsymbol{\theta}_-\mathbf{V}_0 - \mathbf{V}_1 \\ \\ -\boldsymbol{\theta}_+\mathbf{V}_0 + \mathbf{V}_1 \end{pmatrix}, \tag{6.4.3}$$

where $\Delta\mathbf{L} \equiv -\omega^2 mQ\mathbf{I}/K$, and $\boldsymbol{\theta}_\pm$ are the eigenvalues of **T** given by (1.6.2). Here we denoted the matrix $(\boldsymbol{\theta}_- - \boldsymbol{\theta}_+)^{-1}$ by $1/(\boldsymbol{\theta}_- - \boldsymbol{\theta}_+)$. Equation (6.4.3) means, according to (1.5.3) and (1.5.4), that there must exist a vector **z** such that

$$\left[\boldsymbol{\theta}_+^N - \boldsymbol{\theta}_-^N - \Delta\mathbf{L}\left\{\frac{\boldsymbol{\theta}_+^N}{(\mathbf{I} - \boldsymbol{\theta}_+)} - \frac{\boldsymbol{\theta}_-^N}{(\mathbf{I} - \boldsymbol{\theta}_-)}\right\}\right]\mathbf{z} = \mathbf{0}. \tag{6.4.4}$$

This in turn requires that

$$\left|\boldsymbol{\theta}_+^N - \boldsymbol{\theta}_-^N - \Delta\mathbf{L}\left\{\frac{\boldsymbol{\theta}_+^N}{(\mathbf{I} - \boldsymbol{\theta}_+)} - \frac{\boldsymbol{\theta}_-^N}{(\mathbf{I} - \boldsymbol{\theta}_-)}\right\}\right| = 0. \tag{6.4.5}$$

By using (6.1.4) and (6.1.9), this can be reduced to

$$\prod_{\mu=0}^{M-1} \{\tan\beta_\mu \sin 2N\beta_\mu\}\left\{1 + \frac{\omega^2 mQ}{2K}(1 + \cot\beta_\mu \cot 2N\beta_\mu)\right\} = 0, \tag{6.4.6}$$

which was first derived by Asahi (1964) and Asahi and Hori (1963).

For the case $Q = 0$, i.e. for the regular lattice, the roots of (6.4.6) are

$$\beta_\mu = \varrho\pi/2N, \quad \varrho = 1, 2, \ldots, N, \tag{6.4.7}$$

so that the eigenfrequencies are given by

$$m\omega_{\varrho\mu}^2 = 4K\sin^2(\varrho\pi/2N) + 4K'\sin^2(\mu\pi/2M),$$
$$\varrho = 1, 2, \ldots, N; \quad \mu = 1, 2, \ldots, M. \tag{6.4.8}$$

This coincides with formula (1.6.8), as it should.

For $Q \neq 0$, the first factors of (6.4.6) cannot give any root, since each of its zeros is cancelled by the corresponding pole of the second factor. The eigenfrequencies are, therefore, given by the roots of the second factor, some of which may be complex. For the complex roots $\beta_\mu = (\pi/2) + i\varepsilon_\mu$, we have

$$1 + \omega^2 mQ/2K \exp \varepsilon_\mu \cosh \varepsilon_\mu = 0, \tag{6.4.9}$$

in the limit $N \to \infty$. This just coincides with (6.2.8).

In the one-dimensional case, (6.4.6) reduces to

$$\tan \beta \sin 2N\beta \{1 + \omega^2 mQ(1 + \cot \beta \cot 2N\beta)/2K\} = 0 \tag{6.4.10}$$

or

$$\tan (2N - 1)\beta + (1 + 2Q) \tan \beta = 0. \tag{6.4.11}$$

This formula was first obtained by Matsuda (1962b) and later by Asahi (1964). From this it is seen that the effect of the impurity is to decrease or increase the eigenfrequencies according as $Q > 0$ or $Q < 0$. The values of the perturbed eigenfrequencies cannot, however, exceed those of neighbouring unperturbed ones. When $Q < -\frac{1}{2}$, the maximum eigenfrequency runs out from the frequency band. For this impurity frequency, (6.4.11) becomes just (6.1.31).

The spatial variation of the displacement in each normal mode is easily calculated from the equation

$$\mathbf{X}_n = \mathbf{T}^{N-n}\mathbf{T}'^{-1}\mathbf{X}_{N+1}. \tag{6.4.12}$$

In the one-dimensional case the vector $\mathbf{X}_{N+1}$ has the form $(v_N, v_N)^T$ with an arbitrary v_N, corresponding to the free boundary condition. Then it comes out that the normal modes associated with the in-band frequencies are sinusoidal, whereas that associated with the impurity frequency is localized as

$$v_n \cong (-1)^{N-n} \exp \{-2(N - n)\varepsilon\} v_N. \tag{6.4.13}$$

The argument runs in a similar way also for the two-dimensional case, leading to the conclusion that the normal modes associated with the real roots of (6.4.6) are extended, while those associated with the complex roots are localized in the x-direction, in harmony with the result already obtained in § 6.2.

Asahi (1964) also treated the three-dimensional systems by the above method. For a cubic lattice composed of NML particles situated at the integral points $x = n$, $y = m$, and $z = l(1 \leqq n \leqq N$, $1 \leqq m \leqq M$, $1 \leqq l \leqq L)$, in which the particle at $m = j$, $l = k$ on the face $n = N$ is an impurity with mass m', he obtained the eigenfrequency equation

$$1+\frac{\omega^2 mQ}{2K}\left[1+\frac{1}{ML}\sum_{\mu=0}^{2M-1}\sum_{\nu=0}^{2L-1}\cos^2\left\{\frac{(2j-1)\mu\pi}{2M}\right\}\right.$$

$$\left.\times\cos^2\left\{\frac{(2k-1)\nu\pi}{2L}\right\}\cot\beta_{\mu\nu}\cot 2N\beta_{\mu\nu}\right]=0, \qquad (6.4.14)$$

where

$$m\omega^2\equiv 4K\sin^2\beta_{\mu\nu}+4K'\sin^2(\mu\pi/2M)+4K''\sin^2(\nu\pi/2L), \qquad (6.4.15)$$

K'' being the second non-central force constant associated with the z-direction. For the impurity frequency, (6.4.14) becomes

$$1+\frac{\omega^2 mQ}{2K}\left[1-\frac{1}{ML}\sum_{\mu=0}^{2M-1}\sum_{\nu=0}^{2L-1}\cos^2\left\{\frac{(2j-1)\mu\pi}{2M}\right\}\right.$$

$$\left.\times\cos^2\left\{\frac{(2k-1)\nu\pi}{2L}\right\}\tanh\varepsilon_{\mu\nu}\right]=0, \qquad (6.4.16)$$

where

$$\beta_{\mu\nu}\equiv(\pi/2)+i\varepsilon_{\mu\nu}. \qquad (6.4.17)$$

It is seen that (6.4.16) is the direct generalization of (6.1.33) to the three-dimensional case. In a way similar to the above it can be shown that there exists only one impurity frequency, and each of the other eigenfrequencies becomes larger or smaller than the corresponding unperturbed one, according as $Q<0$ or $Q>0$, but never exceeds the neighbouring unperturbed frequency.

If we put $N=1$ and

$$m\omega_{\mu\nu}^2\equiv 4K'\sin^2(\mu\pi/2M)+4K''\sin^2(\nu\pi/2L), \qquad (6.4.18)$$

in (6.4.14), this turns out to be

$$1+\frac{\omega^2 Q}{ML}\sum_{\mu=0}^{2M-1}\sum_{\nu=0}^{2L-1}\frac{\cos^2\left\{\frac{(2j-1)\mu\pi}{2M}\right\}\cos^2\left\{\frac{(2k-1)\nu\pi}{2L}\right\}}{\omega^2-\omega_{\mu\nu}^2}. \qquad (6.4.19)$$

This is none other than the eigenfrequency equation for the two-dimensional square lattice on $(y-z)$ plane in which the atom at $y=j$, $z=k$ is an isotopic impurity. Thus the eigenfrequency equation for the lattice containing an impurity inside it can be deduced from the equation for the lattice with an impurity on one of its faces and with one more dimension. It should further be noted that (6.4.19) has the form which is familiar in the ordinary Green-function theory of the impurity frequency. This shows that in the problem of isolated impurity, the Green-function theory is as fruitful as the transfer matrix formalism. Actually, many investigations have been done on such problems upon the basis of Green-function formalism, We do not give an account of these works, however, since it is outside the scope of the present book and, moreover, several comprehensive articles have already been published on the Green-function treatments (see the Bibliographical Notes).

The method of transfer matrix was first introduced by Saxon and Hutner (1949) and used by Luttinger (1951), Kerner (1954, 1956), and Schmidt (1957) for the investigation of the electronic spectrum of some disordered systems. Hori and Asahi (1957), Hori (1957), and Schmidt (1957) applied it for the first time to the problem of vibration of disordered chains. Afterwards it was used by Maradudin and Weiss (1958) for calculating the vibrational spectrum of the generalized diatomic chains, by Hori (1961) for discussing the wave form of the normal modes of linear chains, and by Hatakeyama (1965), Kuliasko (1964a, b) and de Dycker and Phariseau (1965) for investigating the electronic spectrum of disordered arrays of potentials. Matsuda (1962a) considerably generalized the method to apply it to the polymer molecules having complex structures resembling the two-dimensional system. The application to the genuine higher-dimensional systems was first made by Asahi (1962, 1964). The transfer matrix formalism is particularly powerful for examining the detailed wave form of the normal modes, for the successive transfer of the state vector just reproduces the spatial variation of the displacement. It involves some excessive calculations, however, for the purpose of discussing the impurity modes only. As has been seen thus far, the phase theory is much simpler and more effective for the problem of impurity modes and the spectral properties of disordered systems.

6.5. Method of Scattering Matrix

For the investigation of impurity modes, we have another method which utilizes the concept of the *scattering matrix* (*S-matrix*). It does not require an excessive calculation as the transfer matrix method, so that it ranks with the phase theory in its simplicity.

As an example consider again the two-dimensional lattice with an impurity edge as treated in §§ 6.2 and 6.4. The equations of motion for the impurity atoms are

$$-m'\omega^2 v_{Nm} = K(v_{N-1,m} - v_{Nm}) + K'(v_{N,m-1} - 2v_{Nm} + v_{N,m+1}),$$
$$m = 1, 2, \ldots, M. \tag{6.5.1}$$

Let us develop the displacement v_{nm} in terms of the eigenfunctions (1.5.21):

$$v_{nm} = \sum_{\mu=0}^{M-1} c_n^{(\mu)} Z_m^{(\mu)}. \tag{6.5.2}$$

Assuming the solution of the equation of motion to be of the form

$$c_n^{(\mu)} = I_\mu \exp(i\beta_\mu n) + R_\mu \exp(-i\beta_\mu n), \tag{6.5.3}$$

and putting it into the equation of motion for the regular part of the lattice, we get the relation (1.5.24). On the right-hand side of (6.5.3), the first and second terms correspond respectively to the incident and reflected (or scattered) waves with respect to the impurity edge. Substituting (6.5.2) in (6.5.1), we have

$$-m'\omega^2 \sum_{\mu=0}^{M-1} c_N^{(\mu)} Z_m^{(\mu)} = K \sum_{\mu=0}^{M-1} \{c_{N-1}^{(\mu)} - c_N^{(\mu)}\} Z_m^{(\mu)}$$
$$- 4K' \sum_{\mu=0}^{M-1} c_N^{(\mu)} Z_m^{(\mu)} \sin^2(\mu\pi/2M), \tag{6.5.4}$$

which becomes

$$-m'\omega^2 c_N^{(\mu)} = K\{c_{N-1}^{(\mu)} - c_N^{(\mu)}\} - 4K' c_N^{(\mu)} \sin^2(\mu\pi/2M), \tag{6.5.5}$$

where we used the orthogonality relation for $Z_m^{(\mu)}$.

If we substitute (6.5.3) in (6.5.5), we get the relation which should hold between the amplitudes of the incident and reflected waves. Since the surface modes can be regarded as the standing waves which persist even when there is no incident wave, the condition for the surface mode to exist is that R_μ remains non-vanishing even if one puts $I_\mu = 0$ in this relation. After putting $I_\mu = 0$, this becomes

$$R_\mu\{-m'\omega^2 \exp(-2i\beta_\mu N) - K \exp(-2i\beta_\mu(N-1))$$
$$+ K \exp(-2i\beta_\mu N) + 4K' \exp(-2i\beta_\mu N) \sin^2(\mu\pi/2M)\} = 0. \tag{6.5.6}$$

According to the above requirement, the expression in the bracket should vanish. Using (1.5.24) and putting $\beta_\mu = (\pi/2) + i\varepsilon_\mu$, we see that this condition just coincides with the formula (6.2.8). From (6.5.3) it is clear that the wave corresponding to the root of (6.2.8) decays exponentially as we recede from the impurity edge.

Next consider a square lattice with one impurity at $m = j$ on the edge $n = N$. The development (6.5.2) and the relation (1.5.24) remain the same as before. The equations of motion for atoms on the edge are

$$-m(1+Q)\,\omega^2 \sum_{\mu=0}^{M-1} c_N^{(\mu)} Z_m^{(\mu)} = K \sum_{\mu=0}^{M-1} \{c_{N-1}^{(\mu)} - c_N^{(\mu)}\}\, Z_m^{(\mu)}$$

$$-4K' \sum_{\mu=0}^{M-1} c_N^{(\mu)} Z_m^{(\mu)} \sin^2(\mu\pi/2M), \quad m = j,$$

$$-m\omega^2 \sum_{\mu=0}^{M-1} c_N^{(\mu)} Z_m^{(\mu)} = K \sum_{\mu=0}^{M-1} c\{_{N-1}^{(\mu)} - c_N^{(\mu)}\}\, Z_m^{(\mu)}$$

$$-4K' \sum_{\mu=0}^{M-1} c_N^{(\mu)} Z_m^{(\mu)} \sin^2(\mu\pi/2M), \quad m \neq j. \tag{6.5.7}$$

By the use of the orthogonality relation for $Z_m^{(\mu)}$, these equations can be simplified to

$$-m\omega^2 c_N^{(\mu)} - \frac{\omega^2 mQ}{M} \cos\left\{\frac{(2j-1)\,\mu\pi}{2M}\right\}$$

$$\times \sum_{\nu=0}^{M-1} (2-\delta_{\nu 0})\, c_N^{(\nu)} \cos\left\{\frac{(2j-1)\,\nu\pi}{2M}\right\}$$

$$= K\{c_{N-1}^{(\mu)} - c_N^{(\mu)}\} - 4K' c_N^{(\mu)} \sin^2\left(\frac{\mu\pi}{2M}\right). \tag{6.5.8}$$

Assuming as before the solution of the form (6.5.3) with $I_\mu = 0$ and substituting it in (6.5.8), we get

$$-\frac{m\omega^2}{K} R_\mu \exp(-2i\beta_\mu N) - \frac{\omega^2 mQ}{MK} \cos\left\{\frac{(2j-1)\,\mu\pi}{2M}\right\}$$

$$\times \sum_{\nu=0}^{M-1} (2-\delta_{\nu 0})\, R_\nu \exp(-2i\beta_\nu N) \cos\left\{\frac{(2j-1)\,\nu\pi}{2M}\right\}$$

$$= \{R_\mu \exp(-2i\beta_\mu(N-1)) - R_\mu \exp(-2i\beta_\mu N)\}$$

$$-4\frac{K'}{K} \sin^2\left(\frac{\mu\pi}{2M}\right) R_\mu \exp(-2i\beta_\mu N). \tag{6.5.9}$$

As the condition for the impurity mode, we require that the determinant of the matrix $\mathbf{S}$ of the coefficients of R_μ's should vanish. After some calculations it turns out to be

$$|\mathbf{S}| = \prod_{\mu=0}^{M-1} \{-2iK \exp(-2i\beta_\mu N) \sin \beta_\mu \exp(i\beta_\mu)\}$$
$$- \frac{\omega^2 mQ}{M} \sum_{\mu=0}^{M-1} (2 - \delta_{\mu 0}) \cos^2 \left\{\frac{(2j-1)\mu\pi}{2M}\right\} \exp(2i\beta_\mu N)$$
$$\times \prod_{\nu \neq \mu} \{-2iK \exp(-2i\beta_\nu N) \sin \beta_\nu \exp(i\beta_\nu)\} = 0. \qquad (6.5.10)$$

When we put $\beta_\mu = (\pi/2) + i\varepsilon_\mu$, the factors of the first product do not vanish for any ε_μ, so that we may divide both sides of the equation by its first term of the left-hand side. Then it just reduces to (6.1.33).

The matrix $\mathbf{S}^{-1}$ is called the scattering matrix since it gives the vector $(R_1, R_2, \ldots, R_M)^T$ representing the scattered wave when it is operated on the vector $(I_1, I_2, \ldots, I_M)$ representing the incident wave. The condition $|\mathbf{S}| = 0$ is just the condition for the poles of the determinant of the *S*-matrix. In the first example $\mathbf{S}^{-1}$ was a diagonal matrix.

S-matrix was for the first time used by Saxon and Hutner (1949) in their discussion on the electronic state of a one-dimensional chain of potentials. It was introduced into the investigation of vibrational problems by Fukuda (1962a) for one-dimensional case, and later generalized for higher-dimensional cases by Hori and Asahi (1964). Osawa and Kotera (1966) presented a variant of the *S*-matrix method, which is intimately related to the method of invariant imbedding often used in the theoretical study of transport phenomena. In practice, the *S*-matrix formalism is merely another version of the phase theory. However, theoretically it has an outstanding significance in that it provides us with a link between the theory of the spectrum and that of the scattering process, as was discussed in detail by Asahi (1966).

6.6. A Saxon–Hutner-type Statement for Higher-dimensional Lattices

We have seen in §§ 6.1–6.3 that the phase theory works as well in higher dimensions with respect to the problems of impurity modes and peak structure. As for the question concerning the spectral gaps in disordered systems, however, it encounters a serious difficulty, since in higher-dimensional cases the phase and the associated quantities become matrices, so that it is not easy to visualize the concepts such as sink-phase interval and locking of the phase, and so on. Fortunately, such a perplexity can be avoided by utilizing the principles of the continued-fraction theory, which have already been described in § 3.6. Matsuda and Okada (1965) showed that in certain particular cases we may derive the statement similar to the theorem of Worpitzky also in higher-dimensional cases by using the concept of bound of the matrix, and deduce therefrom a Saxon–Hutner-type statement.

For a $M \times M$ matrix **A**, the bound is defined by

$$\langle \mathbf{A} \rangle = \max_{\mathbf{x}} N(\mathbf{Ax})/N(\mathbf{x}), \tag{6.6.1}$$

where **x** is an arbitrary M-dimensional column vector and

$$N(\mathbf{x}) \equiv (\mathbf{x}^T\mathbf{x})^{1/2}. \tag{6.6.2}$$

If **A** is Hermitian, $\langle \mathbf{A} \rangle$ is, of course, equal to the largest eigenvalue of **A**. For the bound, it is known that

$$\langle \mathbf{A} + \mathbf{B} \rangle \leqq \langle \mathbf{A} \rangle + \langle \mathbf{B} \rangle \tag{6.6.3}$$

and

$$\langle \mathbf{AB} \rangle \leqq \langle \mathbf{A} \rangle \langle \mathbf{B} \rangle. \tag{6.6.4}$$

If $\langle \mathbf{A} \rangle < 1$, we have

$$\langle (\mathbf{I} + \mathbf{A})^{-1} \rangle \leqq 1/(1 - \langle \mathbf{A} \rangle); \tag{6.6.5}$$

for from the identity

$$(\mathbf{I} + \mathbf{A})^{-1} = \mathbf{I} - \mathbf{A}(\mathbf{I} + \mathbf{A})^{-1}$$

we get

$$\langle (\mathbf{I} + \mathbf{A})^{-1} \rangle \leqq 1 + \langle \mathbf{A} \rangle \langle (\mathbf{I} + \mathbf{A})^{-1} \rangle,$$

from which (6.6.5) follows.

According to (3.6.15), we must have

$$\mathbf{U}(j, l) = \frac{\mathbf{a}_j \mathbf{b}_j^{-1}}{\mathbf{I} + \mathbf{U}(j+1, l)} \mathbf{b}_{j-1}^{-1}, \quad 1 \leqq j \leqq l - 1, \tag{6.6.6}$$

and

$$\mathbf{U}(l, l) = \mathbf{a}_l \mathbf{b}_l^{-1} \mathbf{b}_{l-1}^{-1}. \tag{6.6.7}$$

Therefore, by (6.6.4),

$$\langle \mathbf{U}(j, l) \rangle = \langle \mathbf{a}_j \mathbf{b}_j^{-1} \rangle \langle \{\mathbf{I} + \mathbf{U}(j+1, l)\}^{-1} \rangle \langle \mathbf{b}_{j-1}^{-1} \rangle. \tag{6.6.8}$$

If $\langle \mathbf{U}(j+1, l) \rangle < 1$, the inequality (6.6.5) is valid, so that

$$\langle \mathbf{U}(j, l) \rangle \leqq \{\langle \mathbf{a}_j \mathbf{b}_j^{-1} \rangle \langle \mathbf{b}_{j-1}^{-1} \rangle\}/\{1 - \langle \mathbf{U}(j+1, l) \rangle\}, \tag{6.6.9}$$

from which follows

$$|\langle \mathbf{U}(j, l)\rangle| \leqq \tfrac{1}{2}, \tag{6.6.10}$$

provided that

$$|\varrho_k| \equiv |\langle \mathbf{a}_j\mathbf{b}_j^{-1}\rangle \langle \mathbf{b}_{j-1}^{-1}\rangle| \leqq \tfrac{1}{4}, \quad j = 1, 2, \ldots, \tag{6.6.11}$$

and

$$\langle \mathbf{U}(j+1, l)\rangle \leqq \tfrac{1}{2}.$$

On the other hand,

$$\langle \mathbf{U}(l, l)\rangle \leqq \langle \mathbf{a}_l\mathbf{b}_l^{-1}\rangle \langle \mathbf{b}_{l-1}^{-1}\rangle \leqq \tfrac{1}{4} < \tfrac{1}{2}, \tag{6.6.12}$$

in virtue of (6.6.4). Thus by mathematical induction, we can conclude that

$$\langle \mathbf{U}(1, l)\rangle \leqq \tfrac{1}{2}, \tag{6.6.13}$$

which shows that we may use the inequality (6.6.5) for the matrix $\mathbf{U}(1, l)$. Then by a repeated use of (6.6.6) and the theorem of Worpitzky it can be shown that $\langle(\mathbf{I} + \mathbf{U})^{-1}\rangle$ must be bounded under the condition (6.6.11). Further, from (6.6.13) and (6.6.5), we obtain

$$\langle\{\mathbf{I} + \mathbf{U}(1, l)\}^{-1}\rangle \leqq 1/\{1 - \langle \mathbf{U}(1, l)\rangle\} \leqq 2. \tag{6.6.14}$$

Now, if the determinant of a matrix $\mathbf{A}$ tends to zero, that is, if

$$\det \mathbf{A} = \varepsilon, \quad \varepsilon \to 0, \tag{6.6.15}$$

then at least one of the matrix elements of $\mathbf{A}^{-1}$ must become infinity for $\varepsilon \to 0$. Let $(\mathbf{A}^{-1})_{ij}$ be such an element, then from

$$N(\mathbf{A}^{-1}\mathbf{x}) \geqq |(\mathbf{A}^{-1})_{ij}x_j|, \tag{6.6.16}$$

we obtain

$$\langle \mathbf{A}^{-1}\rangle \geqq |(\mathbf{A}^{-1})_{ij}| \to \infty.$$

Therefore, if $\det\{\mathbf{I} + \mathbf{U}(1, l)\} \to 0$, $\langle\{\mathbf{I} + \mathbf{U}(1, l)\}^{-1}\rangle$ cannot be finite, which contradicts (6.6.14). Thus we must have

$$\det\{\mathbf{I} + \mathbf{U}(1, l)\} \neq 0. \tag{6.6.17}$$

By the assumption (3.6.16) this means that the condition (3.6.8) cannot be fulfilled. Hence we may conclude that if the inequalities (6.6.11) are satisfied throughout an interval $\lambda_1 < \lambda < \lambda_2$, this interval lies in a gap of the mixed system.

As an application of the above statement, let us consider the system for which

$$\mathbf{D}_k = \mathbf{I}, \quad k = 2, 3, \ldots, N. \tag{6.6.18}$$

Putting $n(k) = k + 1$, we obtain

$$\mathbf{b}_k = -\mathbf{C}_{k+1}, \quad \mathbf{a}_k = -\mathbf{I}, \tag{6.6.19}$$

and

$$\varrho_k = -\langle \mathbf{C}_{k+1}^{-1} \rangle \langle \mathbf{C}_k^{-1} \rangle. \tag{6.6.20}$$

Suppose now that there are S kinds of matrices $\mathbf{C}_k$, denoted by $\mathbf{C}^{(i)}$ ($i = 1, 2, \ldots, S$), and consider the corresponding S constituent lattices, in each of which $\mathbf{C}_k = \mathbf{C}^{(i)}$ for all k. For the ith constituent lattice, $\mathbf{U}(1, l)$ is diagonal in the representation which diagonalizes $\mathbf{C}^{(i)}$, so that it is necessary for $\mathbf{U}(1, l)$ to converge that all the eigenvalues of $\mathbf{C}^{(i)}$ are larger than 2. In other words, if an interval (λ_1, λ_2) lies in a gap, all the eigenvalues of $\mathbf{C}^{(i)}$ must be larger than 2 in this interval. If this is the case for every consituent lattice, $|\varrho_k|$ is smaller than 1/4 by (6.6.20), so that the interval lies also in a gap of the mixed lattice. This is the statement of the Saxon–Hutner type established by Matsuda and Okada.

6.7. Special Frequencies in Higher-dimensional Lattices

In this section let us investigate whether the special frequencies exist or not in multi-dimensional lattices. Consider an isotopically disordered square lattice as a typical example. Such a lattice is of the type considered in the previous section, since in this case the matrix $\mathbf{D}_k$ is a constant matrix $K\mathbf{I}$, and one can make it unity by dividing the equation of motion throughout by K. One can therefore attack the problem whether there can exist special frequencies or not in the square lattice. Each constituent lattice is composed of columns (lines along y-direction), all of which have the same arrangement of atoms. If we interchange the row and column, such a lattice becomes the one in which each column is a regular chain composed exclusively of light or heavy atoms, and such two kinds of regular columns are arranged in a disordered way in the row direction. In this case the matrices $\mathbf{C}_k$, $\mathbf{D}_k$, and $\mathbf{L}_k$ become

$$\mathbf{C}_k = \begin{pmatrix} 2(K' + K) - \lambda m_k & -K & & 0 \\ -K & 2(K' + K) - \lambda m_k & & \\ & & \ddots & \\ 0 & & -K & 2(K' + K) - \lambda m_k \end{pmatrix}, \tag{6.7.1}$$

$$\mathbf{D}_k = -K'\mathbf{I}, \tag{6.7.2}$$

and

$$\mathbf{L}_k = -\mathbf{D}_k^{T^{-1}}\mathbf{C}_k =$$

$$\begin{bmatrix} 2\left(\frac{K}{K'}+1\right)-\lambda\frac{m_k}{K'} & -\frac{K}{K'} & & \\ -\frac{K}{K'} & 2\left(\frac{K}{K'}+1\right)-\lambda\frac{m_k}{K'} & & 0 \\ & 0 & \ddots & \\ & & & 2\left(\frac{K}{K'}+1\right)-\lambda\frac{m_k}{K'} \end{bmatrix}, \tag{6.7.3}$$

respectively.

According to (1.5.17) and (1.5.18), the eigenvalues and eigenvectors of $\mathbf{L}_k$ are given by

$$2\left(\frac{K}{K'}+1\right)-\lambda\frac{m_k}{K'}-\frac{2K}{K'}\cos 2\theta_\mu, \quad \theta_\mu \equiv \frac{\mu\pi}{2(N+1)}, \tag{6.7.4}$$

and

$$Z_n^{(\mu)} = \left(\frac{2}{N+1}\right)^{1/2}\sin 2n\theta_\mu, \tag{6.7.5}$$

respectively. If we put, as in § 1.5,

$$\lambda = (4K'/m^{(0)})\sin^2\beta_\mu + (4K/m^{(0)})\sin^2\theta_\mu, \tag{6.7.6}$$

the eigenvalues of $\mathbf{L}^{(0)}$ and $\mathbf{L}^{(1)}$, the regular submatrices corresponding to the columns composed of light and heavy atoms, become simply

$$2\cos 2\beta_\mu \tag{6.7.7}$$

and

$$2\cos 2\beta_\mu - \lambda m^{(0)}Q^{(1)}/K' \tag{6.7.8}$$

respectively, where $m^{(0)}$ and $m^{(1)}$ are the masses of light and heavy atoms respectively, and $Q^{(1)} \equiv (m^{(1)}/m^{(0)}) - 1$.

The eigenvalues of the matrix

$$\mathbf{T}^{(0)} = \begin{pmatrix} \mathbf{L}^{(0)} & -\mathbf{I} \\ \mathbf{I} & \mathbf{0} \end{pmatrix} \tag{6.7.9}$$

are given by

$$\boldsymbol{\theta}_{\pm} = \tfrac{1}{2}\{\mathbf{L}^{(0)} \pm (\mathbf{L}^{(0)2} - 4\mathbf{I})^{1/2}\}. \tag{6.7.10}$$

Since $\mathbf{L}^{(0)}$ commutes with $\mathbf{L}^{(1)}$, all the matrix elements of $\mathbf{T}^{(0)}$ and

$$\mathbf{T}^{(1)} \equiv \begin{pmatrix} \mathbf{L}^{(1)} & -\mathbf{I} \\ \mathbf{I} & \mathbf{0} \end{pmatrix} \tag{6.7.11}$$

commute with one another, and we may treat them as if they were ordinary numbers. Therefore the subsequent calculations run in completely the same way as in § 5.5, so that it comes out that

$$|\varrho_k| = \frac{|\det \mathbf{T}^{(1)}|}{|\text{trace}\, \mathbf{T}^{(1)}\mathbf{T}^{(0)p(k)-1}| \,.\, |\text{trace}\, \mathbf{T}^{(1)}\mathbf{T}^{(0)p(k)-2}|}. \tag{6.7.12}$$

Since the diagonal elements of $\mathbf{T}^{(1)}$ in the representation which diagonalizes $\mathbf{T}^{(0)}$ are given by

$$\boldsymbol{\theta}_{\pm} \mp \boldsymbol{\theta}_{\pm}\, \Delta\mathbf{L}/(\boldsymbol{\theta}_{-} - \boldsymbol{\theta}_{+}), \tag{6.7.13}$$

where

$$\Delta\mathbf{L} \equiv \mathbf{L}^{(1)} - \mathbf{L}^{(0)} = Q^{(1)}m^{(0)}\lambda\mathbf{I}/K', \tag{6.7.14}$$

we have

$$\text{trace}\; \mathbf{T}^{(1)}\mathbf{T}^{(0)p-1} = \boldsymbol{\theta}_{-}^{p} + \boldsymbol{\theta}_{+}^{p} + \frac{(\boldsymbol{\theta}_{-}^{p} - \boldsymbol{\theta}_{+}^{p})\,\Delta\mathbf{L}}{\boldsymbol{\theta}_{-} - \boldsymbol{\theta}_{+}}. \tag{6.7.15}$$

By the commutability of the matrices appearing in the above expressions, we may deal with each diagonal element separately. The diagonal elements of the right-hand side of (6.7.15) are

$$d_{p-1}^{(\mu)} = 2\cos 2p\beta_{\mu} - Q^{(1)}m^{(0)}\lambda \sin 2p\beta_{\mu}/K' \sin 2\beta_{\mu}. \tag{6.7.16}$$

The condition for the convergence of the continued fraction is that

$$|d_p^{(\mu)}| \geqq 2, \quad p = 0, 1, 2, \ldots, \quad s-1; \quad \mu = 1, 2, \ldots, N, \tag{6.7.17}$$

which becomes, as before,

$$Q^{(1)}m^{(0)}\lambda/2K' > \sigma_c^{-}(s, \beta_{\mu}), \quad \mu = 1, 2, \ldots, N, \tag{6.7.18}$$

where $s-1$ is the maximum length of the succession of columns composed of light atoms.

Thus the condition for the appearance of the gaps and, consequently, the special frequencies has become considerably severer than the one-dimensional case. In particular, for $s \to \infty$, (6.7.18) can be satisfied for a finite $Q^{(1)}$ only if all the real β_{μ}'s are rational multiples of $\pi/2$. Therefore we are driven to the conclusion that the special frequencies do not exist in higher-dimensional lattices. Even when s is finite, the situation does not change as long as N is arbitrarily large, for the distribution of β_μ's is then so dense that there is no hope of (6.7.18) being valid for every real β_μ. A similar consideration was made also by Matsuda (1965), with exactly the same conclusion.

It should be remarked, however, that the special frequency is by definition a point at which the density vanishes whatever the concentration of light atoms and arrangement of atoms may be. As will be shown in the next section, there can exist a kind of special frequency which may be called *generalized special frequency* at which the spectral density begins to vanish when the mass ratio exceeds a critical value, if there is some restrictive condition on the concentration or arrangement of atoms.

6.8. Generalized Special Frequencies

Let us again consider the isotopically disordered square lattice and impose the constraint that the light atom never appears in succession in each of its columns. In other words, we consider the case $s = 2$, in which the spectral gaps are given, according to the theorem of Matsuda and Okada, by the common gaps of the regular lattice of heavy atoms and the lattice in which the columns of light and heavy atoms appear alternatingly. According to (6.7.16), the condition for the gaps of the latter lattice is given by

$$\text{(A)} \qquad \cos 4\beta_\mu - \frac{Q^{(1)}m^{(0)}\omega^2}{K'} \cos 2\beta_\mu > 1, \quad \text{for all } \mu, \tag{6.8.1}$$

$$\text{(B)} \qquad \cos 4\beta_\mu - \frac{Q^{(1)}m^{(0)}\omega^2}{K'} \cos 2\beta_\mu < -1, \quad \text{for all } \mu. \tag{6.8.2}$$

Let us first examine the condition (A). If $\cos 2\beta_\mu > 0$, it becomes

$$2Q^{(1)}\omega^2 < -\frac{4K'}{m^{(0)}} \sin 2\beta_\mu \tan 2\beta_\mu, \tag{6.8.3}$$

which can never be satisfied. If $\cos 2\beta_\mu < 0$, it turns out to be

$$\text{(A}_1\text{)} \qquad 2Q^{(1)}\omega^2 > -\frac{4K'}{m^{(0)}} \sin 2\beta_\mu \tan 2\beta_\mu. \tag{6.8.4}$$

For $\beta = i\gamma_\mu$, (A) becomes

$$(A_2) \qquad 2Q^{(1)}\omega^2 < \frac{4K'}{m^{(0)}} \sinh 2\gamma_\mu \tanh 2\gamma_\mu, \tag{6.8.5}$$

while for $\beta_\mu = (\pi/2) + i\lambda_\mu$, it reduces to

$$2Q^{(1)}\omega^2 > -\frac{4K'}{m^{(0)}} \sinh 2\lambda_\mu \tanh 2\lambda_\mu, \tag{6.8.6}$$

which is always fulfilled.

The condition (B) becomes, when $\cos 2\beta_\mu > 0$,

$$(B_1) \qquad 2Q^{(1)}\omega^2 > \frac{4K'}{m^{(0)}} \cos 2\beta_\mu, \tag{6.8.7}$$

while when $\cos 2\beta_\mu < 0$, it turns out

$$2Q^{(1)}\omega^2 < \frac{4K'}{m^{(0)}} \cos 2\beta_\mu, \tag{6.8.8}$$

which can never be satisfied. For $\beta_\mu = i\gamma_\mu$, (B) becomes

$$(B_2) \qquad 2Q^{(1)}\omega^2 > \frac{4K'}{m^{(0)}} \cosh 2\gamma_\mu, \tag{6.8.9}$$

while for $\beta_\mu = (\pi/2) + i\lambda_\mu$, it reduces to

$$2Q^{(1)}\omega^2 < -\frac{4K'}{m^{(0)}} \cosh 2\lambda_\mu, \tag{6.8.10}$$

which can never be valid.

Next plot the function

$$\omega^2 - \frac{4K}{m^{(0)}} \sin^2 \theta_\mu \tag{6.8.11}$$

as a function of θ_μ and $\{4K'/m^{(0)}\} \sin^2 \beta_\mu$ as a function of β_μ in the same figure (Fig. 6.5), regarding θ_μ and β_μ as continuous variables in the interval $(0, \pi/2)$. For fixed ω^2, the value of β_μ corresponding to a given θ_μ is obtained from Fig. 6.5. It is easily seen that for each given ω^2, the interval of θ_μ may be divided into four regions, such that

(1) in the region indicated by —, (A) can never be satisfied, but (B_1) can be fulfilled for a sufficiently large $Q^{(1)}$;
(2) in the region ———, (B_2) or (A_2) may be satisfied for a sufficiently large $Q^{(1)}$;
(3) in the region ···, (B) can never be fulfilled, but (A_1) can be satisfied for a sufficiently large $Q^{(1)}$, provided the β_μ-value

corresponding to the lower end of the curve ··· is larger than $\pi/4$;

(4) in the region —·—·—, (A) is always satisfied while (B) can never be fulfilled.

Thus the ω^2-value under consideration lies in a gap for a sufficiently large $Q^{(1)}$ if and only if the whole curve of (2.13) is composed of — and — — —, or it is composed of —·—·— and ··· and, moreover, its lower end corresponds to a β_μ-value larger than $\pi/4$.

In case $K = K'$, the first of the above conditions is satisfied between $\omega^2 = 6K/(Q^{(1)} + 1)\, m^{(0)}$ and $\omega^2 = 2K/m^{(0)}$, if $Q^{(1)} > 2$. In other words, this interval gives the gap between the acoustical and optical branches of the two-dimensional lattice in which the column of light atoms and that of heavy atoms appear alternatingly. The disordered lattice under consideration, as well as the lattice in which every column is composed exclusively of light or heavy atoms and the light column never appears in succession, should have a gap below $\omega^2 = 2K/m^{(0)}$, if $Q^{(1)} > 2$ and the maximum frequency $8K/m^{(1)}$ of the regular lattice of heavy atoms is smaller than $\omega^2 = 2K/m^{(0)}$, or $Q^{(1)} > 3$. Thus the frequency $\omega^2 = 2K/m^{(0)}$ gives a generalized special frequency, with the critical mass-parameter $Q^{(1)}_{2c}(2K/m^{(0)}) = 3$, for both of the above-mentioned lattices.

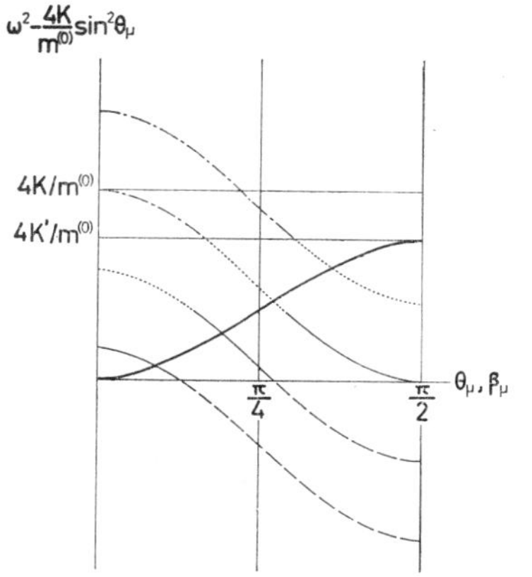

FIG. 6.5. The curves of $\omega^2 - \dfrac{4K}{m^{(0)}} \sin^2 \theta$ as a function of θ for several values of ω^2, and the curve of $\dfrac{4K'}{m^{(0)}} \sin^2 \beta$ as a function of β (thick curve).

The constraint of the type considered above seems, however, to be too weak. For example, in the spectra calculated by Dean and Bacon for an isotopically disordered square lattice with $Q^{(1)} = 2$ (Fig. 2.7), the density seems to vanish at $\omega^2 = 4K/m^{(0)}$ and $5K/m^{(0)}$, when the concentration of light atoms f_L is small. If one recalls that in the one-dimensional case each special frequency corresponds to an *island frequency*, i.e. one of the eigenfrequencies which an appropriate island of light atoms would have if it were imbedded in the sea of infinitely heavy atoms, and considers that the above-mentioned frequencies are just the island frequencies of the islands L and LL, it becomes very likely that they are giving generalized special frequencies under the constraint that f_L is small. The above argument could not predict these frequencies to be generalized special frequencies.

Dean and Bacon (1965) and Payton III and Visscher (1966) conjectured that in multidimensional lattices every island frequency may give a generalized special frequency, if f_L is smaller than the critical percolation probability p_c, so that there exist only the islands of finite size, and consequently there may be only a finite number of island frequencies. Matsuda (1966) showed that this conjecture is indeed true.

Let us consider a dynamical system described by an equation of the form (1.1.20). As is seen from the construction of the matrices $\mathbf{A}_n$ and $\mathbf{B}_n$, the squared-frequencies of such a dynamical system are given by the eigenvalues of the quadratic form

$$H(x, x) = \sum_{j=1}^{\mathrm{N}} \sum_{k=1}^{\mathrm{N}} \sum_{\alpha=1}^{d} \sum_{\beta=1}^{d} \frac{\Phi(j\alpha, k\beta)}{\sqrt{(m_j m_k)}} x_{j\alpha} x_{k\beta}, \tag{6.8.12}$$

where $\Phi(j\alpha, k\beta)$ are the functions of force constants only, j and k indices specifying the equilibrium position of atoms, α and β those specifying the direction of displacement, and d the dimension of the lattice under consideration. The maximum value which $H(x, x)$ may assume under the conditions

$$\sum_{j=1}^{\mathrm{N}} \sum_{\alpha=1}^{d} x_{j\alpha}^2 = 1 \tag{6.8.13}$$

and

$$\sum_{j=1}^{\mathrm{N}} \sum_{\alpha=1}^{d} C(h, j\alpha)\, x_{j\alpha} = 0, \quad h = 1, 2, \ldots, \gamma - 1 \tag{6.8.14}$$

is a function of a set of parameters $C(h, j\alpha)$. As is well known (Courant and Hilbert (1931)), the γth largest squared frequency ω_γ^2

is given by the minimum which this maximum value assumes for a suitable choice of the set $C(h, j\alpha)$.

Suppose now that the mass m_j is either $m^{(0)}$ or $m^{(1)}$ ($m^{(1)} > m^{(0)}$). Then denoting $x_{j\alpha}$ for which $m_j = m^{(0)}$ and $m^{(1)}$ by $y_{j\alpha}$ and $z_{j\alpha}$ respectively, we may write

$$H(x, x) = H^{(0)}(y, y) + H^{(1)}(x, x), \tag{6.8.15}$$

where $H^{(0)}(y, y)$ is the quadratic form of $y_{j\alpha}$'s, and $H^{(1)}(x, x)$ does not contain any term which is quadratic in $y_{j\alpha}$. The eigenvalues of $H^{(0)}(y, y)$ are no other than the island frequencies. Let $\bar{\omega}_p$ be the pth largest island frequency, and $g(p)$ its degree of degeneracy. Then the diagonal form of $H^{(0)}(y, y)$ is

$$H^{(0)}(y, y) = \sum_{p=1}^{r} \bar{\omega}_p{}^2 \sum_{q=1}^{g(p)} \eta_{p,q}^2, \tag{6.8.16}$$

where r is the number of different island frequencies, which is assumed to be finite. If we put $G(\nu) \equiv \sum_{p=1}^{\nu} g(p)$, $G(0) \equiv 0$, we must have $G(r) = dN^{(0)}$, $N^{(0)}$ being the number of light atoms.

Let $H_\nu(x, x)$ be the quadratic form in $dN - G(\nu)$ variables obtained by putting $\eta_{p,q} = 0$ ($p = 1, 2, \ldots, \nu$; $q = 1, 2, \ldots, g(p)$) in $H(x, x)$, and $\omega_\gamma^{(\nu)2}$ be the γth largest eigenvalue of $H_\nu(x, x)$. Further let $G(\nu) \geqq \gamma > G(\nu - 1)$. Then $\bar{\omega}_\nu^2$ is the $G(\nu)$th eigenvalue of $H^{(0)}(y, y)$, and ω_γ^2 is one of the eigenvalues $\omega_{G(\nu-1)+1}^2$, $\omega_{G(\nu-1)+2}^2, \ldots, \omega_{G(\nu)}^2$, so that $\omega_\gamma^2 \geqq \omega_{G(\nu)}^2$. It is clear that $\omega_{G(\nu)}^2 \geqq \bar{\omega}_\nu^2$, since $\omega_{G(\nu)}^2$ and $\bar{\omega}_\nu^2$ are the $G(\nu)$th largest eigenvalues of $H(x, x)$ and $H^{(0)}(y, y)$ respectively ($H^{(0)}(y, y)$ is obtained from $H(x, x)$ by imposing the constraints $z_{j\alpha} = 0$). Thus we have $\omega_\gamma \geqq \bar{\omega}_\nu$. Next consider $\omega_1^{(\nu-1)2}$. This is the maximum eigenvalue of $H_{\nu-1}(x, x)$, or the maximum value of $H(x, x)$ under $G(\nu - 1) + 1$ constraints $\eta_{p,q} = 0$ ($p = 1, 2, \ldots, \nu - 1$; $q = 1, 2, \ldots, g(p)$) and (6.8.13). This is clearly larger than or equal to $\omega_{G(\nu-1)+1}^2$, which is the minimum of the maximum value of $H(x, x)$ under $G(\nu - 1) + 1$ constraints (6.8.13) and (6.8.14). By the same reason we have $\omega_{\gamma-G(\nu-1)}^{(\nu-1)} \geqq \omega_\gamma$. Thus

$$\omega_{\gamma-G(\nu-1)}^{(\nu-1)} \geqq \omega_\gamma \geqq \bar{\omega}_\nu, \quad \bar{\omega}_{r+1} \equiv 0. \tag{6.8.17}$$

Now let $m^{(0)}$ be fixed and suppose that $m^{(1)}$ is sufficiently larger than $m^{(0)}$. Since $H^{(1)}(x, x)$ may then be written

$$|H^{(1)}(x, x)| = \sqrt{\left(\frac{m^{(0)}}{m^{(1)}}\right)} \sum \bar{h}_{j\alpha,\, k\beta} x_{j\alpha} x_{k\beta}, \; \bar{h}_{j\alpha,\, k\beta} = 0(1), \tag{6.8.18}$$

we get an estimation

$$\begin{aligned} |H^{(1)}(x, x)| &\leqq \sqrt{\left(\frac{m^{(0)}}{m^{(1)}}\right)} \max(\bar{h}_{j\alpha,\, k\beta}) \sum |x_{j\alpha} x_{k\beta}| \\ &\leqq \sqrt{\left(\frac{m^{(0)}}{m^{(1)}}\right)} \max(\bar{h}_{j\alpha,\, k\beta}) \cdot \tfrac{1}{2} \sum (|x_{j\alpha}|^2 + |x_{k\beta}|^2) \\ &\leqq \sqrt{\left(\frac{m^{(0)}}{m^{(1)}}\right)} \max(\bar{h}_{j\alpha,\, k\beta}) \sum |x_{j\alpha}|^2 \cdot z, \end{aligned} \tag{6.8.19}$$

where z is the number of neighbouring atoms which are coupled with an atom. Since z is finite, we reach

$$|H^{(1)}(x, x)| = O\left[\sqrt{\left(\frac{m^{(0)}}{m^{(1)}}\right)}\right], \tag{6.8.20}$$

and consequently

$$H(x, x) = H^{(0)}(y, y) + O\left[\sqrt{\left(\frac{m^{(0)}}{m^{(1)}}\right)}\right]. \tag{6.8.21}$$

Since $\omega_1^{(\nu-1)2}$ and $\bar{\omega}_\nu^2$ are the maximum values of $H(x, x)$ and $H^{(0)}(y, y)$ respectively, under the same $G(\nu - 1) + 1$ constraints (6.8.13) and $\eta_{p,q} = 0 \; (p = 1, 2, \ldots, \nu - 1; \; q = 1, \ldots, g(p))$, we have finally

$$\omega^{(\nu-1)2}_{\nu - G(\nu-1)} \leqq \bar{\omega}_\nu^2 + O\left[\sqrt{\left(\frac{m^{(0)}}{m^{(1)}}\right)}\right]. \tag{6.8.22}$$

Therefore, if $m^{(1)}/m^{(0)}$ exceeds a certain critical value and the number of different island frequencies is finite, the interval $(\bar{\omega}_\nu^2 + O(m^{(0)}/m^{(1)}), \bar{\omega}_{\nu-1}^2)$ becomes a spectral gap. This means that every island frequency becomes a generalized special frequency, under the constraint that its total number is finite, when the mass ratio exceeds a certain critical value. Thus it has become certain that the above-mentioned frequencies $\omega^2 = 4K/m^{(0)}$ and $5K/m^{(0)}$, for example, give generalized special frequencies of isotopically disordered square lattice, although the actual values of critical mass ratio have not yet been evaluated.

The constraint considered at the beginning of this section does not prevent the formation of an infinite island, a row composed exclusively of light atoms, which gives rise to a continuum of island frequencies. This is the resaon why the frequencies $\omega^2 = 4K/m^{(0)}$ and $5K/m^{(0)}$ could not be predicted to be generalized special frequencies from the theorem of Matsuda and Okada. The example given there shows, however, that even if there exists a continuous band of island frequencies, its lowest frequency may give a generalized special frequency. In other words, there may appear generalized special frequencies even when $f_L > p_c$, if a very specific constraint is imposed on the arrangement of atoms.

CHAPTER 7

Approximate Theories

7.1. Method of Green Function for Vibrational Spectrum

Among various approximate methods, the most important ones are those derived from the Green function described in § 2.3. Consider a disordered isotopic chain containing two kinds of atoms with masses $m^{(0)}$ and $m^{(1)}(m^{(0)} < m^{(1)})$ and let $f_L(\ll 1)$ be the probability with which the mass $m^{(0)}$ appears. Following Langer (1961), let us write the equation of motion in the form

$$\ddot{v}_n - K(v_{n+1} - 2v_n + v_{n-1})/m^{(1)} = K\{(1/m_n) - (1/m^{(1)})\} \times (v_{n+1} - 2v_n + v_{n-1}), \tag{7.1.1}$$

regarding the regular chain with mass $m^{(1)}$ as the unperturbed system. The transformation to the normal coordinate q_μ of the unperturbed lattice is effected by

$$v_n = \frac{1}{(N\, m^{(1)})^{1/2}} \sum_{\mu=-N/2}^{N/2} q_\mu \exp(2\pi i\mu n/N - i\omega t). \tag{7.1.2}$$

Equation (7.1.1) then becomes

$$(\omega_\mu^2 - \omega^2)q_\mu + \sum_\nu L\mu_\nu q_\nu = 0, \tag{7.1.3}$$

where

$$\omega_\mu = (4K/m^{(1)})^{1/2} \sin(\mu\pi/N) \equiv \omega_{\max} \sin(\mu\pi/N) \tag{7.1.4}$$

are the eigenfrequencies of the unperturbed chain, and

$$L_{\mu\nu} \equiv (\omega^2 \nu/N) \sum_n \{(m^{(1)}/m_n) - 1\} \exp\{2\pi i(\nu - \mu)\, n/N\}. \tag{7.1.5}$$

Equation (7.1.3) has the same form as (2.3.6), if we put $\lambda = \omega^2$. Thus the elements of the inverse of the Green operator is given by

$$\{G^{-1}(\omega^2)\}_{\mu\nu} = (\omega_\mu^2 - \omega^2)\, \delta_{\mu\nu} + L_{\mu\nu}. \tag{7.1.6}$$

According to (2.3.12), the spectral density of the squared frequency ω^2 is given by

$$D(\omega^2) = \lim_{\substack{\varepsilon \to 0 \\ N \to \infty}} \frac{1}{\pi N} \operatorname{trace} G(\omega^2 + i\varepsilon). \tag{7.1.7}$$

Therefore, the density of the frequency itself is

$$D(\omega) = \frac{2\omega}{\pi} \lim_{\substack{\varepsilon \to 0 \\ N \to \infty}} \frac{1}{N} \operatorname{Im} \operatorname{trace} G(\omega^2 + i\varepsilon). \tag{7.1.8}$$

From (7.1.6), it is seen that $G_{\mu\nu}(\omega^2)$ satisfies the equation

$$G_{\mu\nu}(\omega^2) = \frac{\delta_{\mu\nu}}{\omega_\mu^2 - \omega^2} - \frac{1}{\omega_\mu^2 - \omega^2} \sum_\varrho L_{\mu\varrho} G_{\varrho\nu}(\omega^2), \tag{7.1.9}$$

the iteration of which gives

$$\begin{aligned} G_{\mu\nu}(\omega^2) &= \frac{\delta_{\mu\nu}}{\omega_\mu^2 - \omega^2} - \frac{1}{\omega_\mu^2 - \omega^2} L_{\mu\nu} \frac{1}{\omega_\mu^2 - \omega^2} \\ &+ \frac{1}{\omega_\mu^2 - \omega^2} \sum_\varrho \frac{L_{\mu\varrho} L_{\varrho\nu}}{\omega_\varrho^2 - \omega^2} \frac{1}{\omega_\nu^2 - \omega^2} + \dots \end{aligned} \tag{7.1.10}$$

Now we average (7.1.10) over all configurations of atoms. The average of $L_{\mu\nu}$ obviously becomes

$$\langle L_{\mu\nu} \rangle = (f_L Q \omega_\nu^2 / N) \sum_n \exp\{2\pi i(\nu - \mu)\, n/N\} = f_L Q \omega_\nu^2 \delta_{\mu\nu}, \tag{7.1.11}$$

where we have taken $m^{(0)}$ as the standard mass and defined Q by $Q \equiv (m^{(1)}/m^{(0)}) - 1$. Next consider the second-order term

$$\begin{aligned} \langle L_{\mu\varrho} L_{\varrho\nu} \rangle &= \Big\langle (\omega_\varrho^2 \omega_\nu^2 / N^2) \sum_{n,k} \left(\frac{m^{(1)}}{m_n} - 1 \right) \left(\frac{m^{(1)}}{m_k} - 1 \right) \\ &\times \exp\{2\pi i(\varrho - \mu)\, n/N\} \exp\{2\pi i(\nu - \varrho)\, k/N\} \Big\rangle. \end{aligned} \tag{7.1.12}$$

This comes out to be

$$\begin{aligned} \langle L_{\mu\varrho} L_{\varrho\nu} \rangle &= (f_L Q^2 \omega_\varrho^2 \omega_\nu^2 / N^2) \sum_n \exp\{2\pi i(\nu - \mu)\, n/N\} \\ &+ (f_L^2 Q^2 \omega_\varrho^2 \omega_\nu^2 / N^2) \sum_{n \neq k} \exp\{2\pi i(\varrho - \mu)\, n/N\} \exp\{2\pi i(\nu - \varrho)\, k/N\} \\ &= (\omega_\varrho^2 \omega_\nu^2 Q^2 / N)(f_L - f_L^2)\, \delta_{\mu\nu} + \omega_\varrho^2 \omega_\nu^2 Q^2 f_L^2 \delta_{\mu\varrho} \delta_{\varrho\nu}, \end{aligned} \tag{7.1.13}$$

where in performing the summation, the term $n = k$ has been added and then subtracted, producing a correction of order f_L^2 in the coefficient of $\delta_{\mu\nu}$.

The situation is similar also for the general term:

$$\langle L_{\mu(1),\mu(2)} \dots L_{\mu(l),\mu(l+1)} \rangle = (\omega^2_{\mu(2)} \dots \omega^2_{\mu(l+1)}/N^l)$$
$$\times \Big\langle \sum_{n(1),\dots,n(l)} \left(\frac{m^{(1)}}{m_{n(1)}} - 1\right) \dots \left(\frac{m^{(1)}}{m_{n(l)}} - 1\right)$$
$$\times \exp\{2\pi i(p(1)n(1) + \dots + p(i)n(i) + \dots)/N\}\Big\rangle, \qquad (7.1.14)$$

where $p(i)$ is the momentum transfer $(\mu(i+1) - \mu(i))$ in the ith scattering event. Let us write down all possible partitions of the $n(i)$'s, setting $n(i)$'s equal in each group. That is,

$$\langle L_{\mu(1),\mu(2)}, \dots \rangle$$
$$= (\omega^2_{\mu(1)} \dots /N^l) \Big\langle \sum_n \left(\frac{m^{(1)}}{m_n} - 1\right)^l \exp\{2\pi i n(p(1) + \dots + p(l))/N\}\Big\rangle$$
$$+ (\omega^2_{\mu(1)} \dots /N^l) \sum_{s(1)+s(2)=l}^{(2)} \Big\langle \sum_{n(1) \neq n(2)} \left(\frac{m^{(1)}}{m_{n(1)}} - 1\right)^{s(1)} \left(\frac{m^{(1)}}{m_{n(2)}} - 1\right)^{s(2)}$$
$$\times \exp\{2\pi i n(1)(p(1) + \dots)/N\} \exp\{2\pi i n(2)(p(1) + \dots)/N\}\Big\rangle$$
$$+ (\omega^2_{\mu(1)} \dots /N^l) \sum_{s(1)+s(2)+s(3)=l}^{(3)} \Big\langle \sum_{n(1) \neq n(2) \neq n(3)} \left(\frac{m^{(1)}}{m_{n(1)}} - 1\right)^{s(1)}$$
$$\times \left(\frac{m^{(1)}}{m_{n(2)}} - 1\right)^{s(2)} \left(\frac{m^{(1)}}{m_{n(3)}} - 1\right)^{s(3)}$$
$$\times \exp\{2\pi i n(1)(p(1) + \dots)/N\} \dots \Big\rangle + \dots, \qquad (7.1.15)$$

where $\sum^{(2)}, \sum^{(3)}, \dots,$ mean the summation over all partitions of the $n(i)$'s into two, three, ..., groups, respectively. Performing the average and summing over the $n(i)$'s, we find an expansion of the form

$$(\omega^2_{\mu(1)} \dots /N^{l-1}) Q^l P_l(f_L) \delta_{p(1)+\dots+p(l),0}$$
$$+ (\omega^2_{\mu(1)} \dots /N^{l-2}) Q^l \sum_{s(1),s(2)}^{(2)} P_{s(1)}(f_L) P_{s(2)}(f_L)$$
$$\times \delta_{p(1)+\dots,0}\delta_{p(i)+\dots,0} + (\omega^2_{\mu(1)} \dots /N^{l-3}) Q^l$$
$$\times \sum_{s(1),s(2),s(3)}^{(3)} P_{s(1)}(f_L) P_{s(2)}(f_L) P_{s(3)}(f_L) \delta \dots \delta \dots \delta \dots + \dots, \qquad (7.1.16)$$

where $P_s(f_L)$ is an sth-degree polynomial in f_L. For small f_L, the leading terms in P_s are

$$P_s = f_L - (2^{s-1} - 1)f_L^2 + \dots \tag{7.1.17}$$

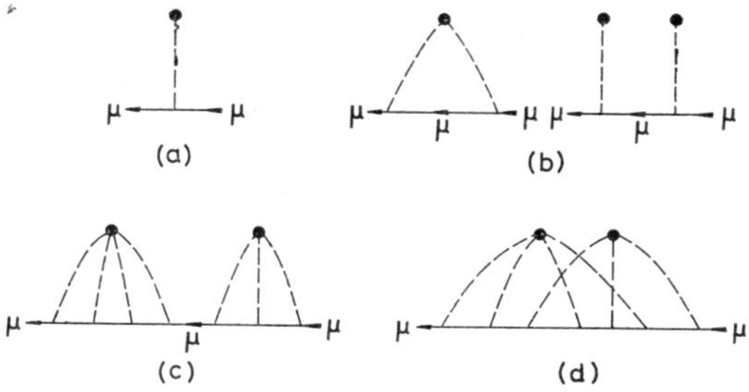

FIG. 7.1. Typical graphs occurring in the perturbation expansion of $G_\mu(\omega^2)$. (After Langer (1961).)

It is useful to represent the various terms occurring in the expansion of $\langle G_{\mu\mu}(\omega^2)\rangle$ diagrammatically. In Fig. 7.1a horizontal line represents the phonon of momentum μ and frequency ω whose propagation is described by $\langle G_{\mu\mu}(\omega^2)\rangle$. The interactions are denoted by broken lines which start at dots representing the lattice points where the interactions occur and are connected with the phonon line in the order in which they occur in the expansion (7.1.10). Figures 7.1a and 7.1b are the graphs associated with (7.1.11) and (7.1.13) respectively. In Figs. 7.1c and 7.1d typical graphs associated with the $\sum^{(2)}$ terms in (7.1.16) are depicted.

The proper self-energy part of a diagram is defined to be the one which cannot be broken into two such parts simply by cutting the phonon line once. For example, Fig. 7.1c contains two proper parts, whereas Fig. 7.1d contains only one. Every proper self-energy part is diagonal in the μ index. Let us denote the sum of all proper self-energy parts by $D_\mu(\omega^2)$. It is known that

$$G_\mu(\omega^2) = \{\omega_\mu^2 - \omega^2 + D_\mu(\omega^2)\}^{-1}, \tag{7.1.18}$$

where $\langle G_{\mu\mu}(\omega^2)\rangle$ is written simply as $G_\mu(\omega^2)$.

According to (7.1.16) and (7.1.17), the only self-energy graphs which contain contributions linear in f_L are those in which all the scattering events occur at the same site. Thus the leading contribution to $D_\mu(\omega^2)$ is the sum of the graphs shown in Fig. 7.2. To the lowest order in f_L, this sum is given by

$$D^{(1)}(\omega^2) = f_L\omega_\mu^2 \sum_{l=1}^{\infty} (-1)^{l+1}Q^l/N^{l-1} \times \sum_{\mu(1),\ldots,\mu(l)} \frac{\omega_{\mu(1)}^2}{\omega_{\mu(1)}^2 - \omega^2} \cdots \frac{\omega_{\mu(l-1)}^2}{\omega_{\mu(l-1)}^2 - \omega^2}$$

$$= f_L Q\omega_\mu^2 \left[1 + \frac{Q}{N}\sum_\nu \frac{\omega_\nu^2}{\omega_\nu^2 - \omega^2}\right]^{-1}$$

$$= f_L Q'\omega_\mu^2 \left[1 + \frac{Q'\omega^2}{N}\sum_\nu \frac{1}{\omega_\nu^2 - \omega^2}\right]^{-1}, \qquad (7.1.19)$$

where $Q' \equiv Q/(1+Q) = 1 - (m^{(0)}/m^{(1)})$.

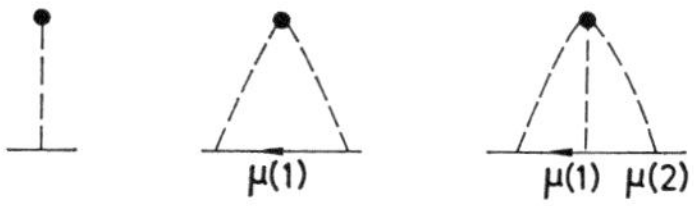

FIG. 7.2. The graphs which contribute to $D_\mu^{(1)}(\omega^2)$. (After Langer (1961).)

The summation in the second expression in (7.1.19) becomes in the limit $N \to \infty$, $\varepsilon \gg \omega_{\max}/N \to 0$,

$$\lim_{\substack{N\to\infty \\ \varepsilon\to 0}} \frac{1}{N}\sum_\mu (\omega_\mu^2 - \omega^2 - i\varepsilon)^{-1} = \int_{-\pi/2}^{\pi/2} dx\{\pi(\omega_{\max}^2 \sin^2 x - \omega^2)\}^{-1}$$

$$= i/\{\omega(\omega_{\max}^2 - \omega^2)^{1/2}\}, \qquad (7.1.20)$$

so that we get

$$G_\mu(\omega^2) = \frac{1}{\omega_\mu^2 - \omega^2 + f_L Q'\omega_\mu^2(\omega_{\max}^2 - \omega^2)^{1/2}/\{(\omega_{\max}^2 - \omega^2)^{1/2} + iQ'\omega^2\}}, \quad \omega^2 \leqq \omega_{\max}^2, \qquad (7.1.21)$$

$$G_\mu(\omega^2) = \frac{1}{\omega_\mu^2 - \omega^2 + f_L Q'\omega_\mu^2(\omega^2 - \omega_{\max}^2)^{1/2}/\{(\omega^2 - \omega_{\max}^2)^{1/2} - Q'\omega^2\}}, \quad \omega^2 > \omega_{\max}^2. \qquad (7.1.22)$$

From (7.1.21) it turns out that $G_\mu(\omega^2)$ has a single pole for each value of μ, which occurs at

$$\omega = \omega_\mu + \varDelta_\mu - i\Gamma_\mu; \quad \varDelta_\mu \equiv \frac{\omega_\mu f_L Q'(\omega_{\max}^2 - \omega_\mu^2)}{2\{\omega_{\max}^2 - \omega_\mu^2(1 - Q'^2)\}},$$

$$\Gamma_\mu \equiv \frac{\omega_\mu^2 f_L Q'^2(\omega_{\max}^2 - \omega_\mu^2)^{1/2}}{2\{\omega_{\max}^2 - \omega_\mu^2(1 - Q'^2)\}}. \qquad (7.1.23)$$

For $\omega^2 > \omega_{\max}^2$ we find from (7.1.22) that it may have a pole on the real axis just above the impurity frequency.

For the spectral density $D(\omega)$ we finally obtain

$$D(\omega) = \frac{2}{\pi}\mathrm{Re}$$
$$\times \left[\frac{1}{\{\omega_{\max}^2 - \omega^2 + \omega_{\max}^2 f_L Q'(\omega_{\max}^2 - \omega^2)^{1/2}/((\omega_{\max}^2 - \omega^2)^{1/2} + iQ'\omega)\}^{1/2}}\right],$$
$$\omega \leqq \omega_{\max}, \qquad (7.1.24\text{a})$$
$$= \frac{2}{\pi}\mathrm{Re}$$
$$\times \left[\frac{1}{\{\omega_{\max}^2 - \omega^2 + \omega_{\max}^2 f_L Q'(\omega^2 - \omega_{\max}^2)^{1/2}/((\omega^2 - \omega_{\max}^2)^{1/2} - Q'\omega)\}^{1/2}}\right],$$
$$\omega > \omega_{\max}. \qquad (7.1.24\text{b})$$

Equation (7.1.24) gives the spectrum as shown by broken curves in Fig. 7.3. For the low-frequency region it coincides with the unperturbed spectrum, while at the upper end of the band, the correction term predominates in (7.1.24a). A singularity still occurs at $\omega^2 = \omega_{\max}^2$, but its strength is reduced from inverse square root to inverse fourth root. There appears an impurity band which extends

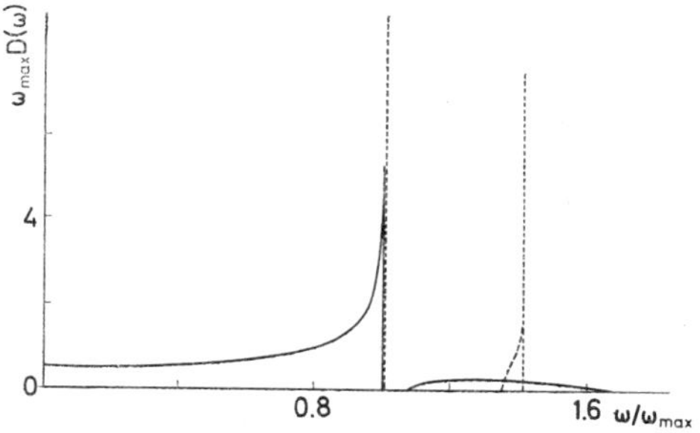

FIG. 7.3. The density of eigenfrequency spectrum of the isotopically disordered diatomic chain for the case $Q^{(1)} = 2$ and $f_L = 0{\cdot}1$. The broken and full curves correspond to the approximation (7.1.24) and the self-consistent formula (7.1.32) respectively. (After Davies and Langer (1963).)

within a small range of frequencies just above the impurity frequency. Comparison with Fig. 2.6a shows, however, that the present calculation fails to yield the fine structure in the impurity band and the presence of other minor peaks outside the main band.

Davies and Langer (1963) made a step forward and constructed a *self-consistent field approximation*, taking into account that the exact expressions for $G_\mu(\omega^2)$ and $D_\mu(\omega^2)$ should have very similar analytic properties. In particular, both functions must be analytic everywhere in the ω-plane except along the real axis where there are branch cuts. These branch cuts occur wherever $D(\omega)$ is non-zero. In the above approximation, $G_\mu(\omega^2)$ and $D_\mu(\omega^2)$ both have branch cuts coinciding with the band of the unperturbed lattice. However, while the pole of $D_\mu^{(1)}(\omega^2)$ outside the band lies at the impurity frequency, the position of the pole of $G_\mu(\omega^2)$ depends upon μ, and $D(\omega)$ is non-zero in a narrow band in this region.

By (7.1.19), $D_\mu^{(1)}(\omega^2)$ can be written

$$D_\mu^{(1)}(\omega^2) = Nf_L t_{\mu\mu}(\omega^2), \tag{7.1.25}$$

where

$$t_{\mu\mu}(\omega^2) \equiv \frac{Q'\omega_\mu^2/N}{1 + (Q'\omega^2/N)\sum_\nu 1/(\omega_\nu^2 - \omega^2)} = \frac{Q\omega_\mu^2/N}{1 + Qf(\omega)},$$

$$f(\omega) \equiv \frac{1}{N}\sum_\nu \frac{\omega_\nu^2}{\omega_\nu^2 - \omega^2}. \tag{7.1.26}$$

Defining

$$t_{\mu\nu}(\omega^2) \equiv \frac{Q\omega_\nu^2/N}{1 + Qf(\omega)}, \tag{7.1.27}$$

we have the equation to be satisfied by $t_{\mu\nu}(\omega^2)$:

$$t_{\mu\nu}(\omega^2) \equiv \frac{Q\omega_\nu^2}{N} - \frac{1}{N}\sum_\varrho \frac{Q\omega_\varrho^2\, t_{\varrho\nu}(\omega^2)}{\omega_\varrho^2 - \omega^2}. \tag{7.1.28}$$

The above requirement of analytic self-consistency can be met by modifying (7.1.28) to

$$t_{\mu\nu}(\omega^2) = \frac{Q\omega_\nu^2}{N} - \frac{1}{N}\sum_\varrho Q\omega_\varrho^2 G_\varrho(\omega^2)\, t_{\varrho\nu}(\omega^2), \tag{7.1.29}$$

which becomes, according to (7.1.18),

$$t_{\mu\nu}(\omega^2) = \frac{Q\omega_\nu^2}{N} - \frac{Q}{N}\sum_\varrho \frac{\omega_\varrho^2 t_{\varrho\nu}(\omega^2)}{\omega_\varrho^2 - \omega^2 + D_\varrho(\omega^2)}. \tag{7.1.30}$$

This can be solved easily, with the result

$$t_{\mu\nu}(\omega^2) = \frac{NQ\omega_\nu^2}{1 + (Q/N) \sum_\varrho [\omega_\varrho^2/\{\omega_\varrho^2 - \omega^2 + D_\varrho(\omega^2)\}]}, \quad (7.1.31)$$

which leads to

$$D_\mu(\omega^2) = \frac{f_L Q\omega_\mu^2}{1 + (Q/N) \sum_\varrho [\omega_\varrho^2/\{\omega_\varrho^2 - \omega^2 + D_\varrho(\omega^2)\}]}, \quad (7.1.32)$$

by (7.1.25). Comparing this with (7.1.26), we see that in (7.1.32) the unperturbed Green function has been replaced by the exact one in each of the phonon lines of the diagrams contributing to (7.1.26). The solution of (7.1.32) is thus equivalent to the summation of a very large class of diagrams.

The spectrum obtained by solving (7.1.32) is plotted by full curves in Fig. 7.3. The main band is identical with that obtained previously except in the immediate neighbourhood of the band edge, while the impurity band has become quite broad and flat. Comparing with Dean's spectrum, we see that the reproduction of fine structures has become rather worse than in the preceding approximation, although the overall distribution of energy seems to be represented better. Thus it is seen that the approximation of the above kind can give only a very poor result, in particular in the high-frequency region, notwithstanding the fact that a large class of diagrams are summed up consistently. The reason for this shortcoming in the method probably lies in that the normal modes are strongly localized in such a region, as was shown in §§ 2.6 and 5.7, since in the above treatment the plane waves are used as the zeroth-order solution, so that it is implicitly assumed from the beginning that such extended waves represent a good approximation.

Upon the basis of a similar method of diagram expansion, Takeno (1960, 1962a, c, 1963) discussed qualitatively the behaviour of the spectrum of three-dimensional cubic lattices containing the impurities in small concentrations, and obtained an important result that very heavy impurities give rise to well-defined resonance modes, which appear in the spectrum as a sharp peak in the low-frequency region. Mozer and Maradudin (1963) calculated the frequency spectrum for an isotopically disordered face-centred cubic lattice by an extension of Langer's method and obtained a similar result. The existence of the resonance modes reveals itself in an anomaly of low-temperature specific heat, a dip in the temperature-dependence of thermal conductivity, and so on, which are experimentally very important. However, as in the one-dimensional case, information concerning finer structure of the spectrum has not yet been obtained by these approximate methods.

7.2. Method of Green Function for Electronic Spectrum

The method of Green function can be applied also to the systems governed by differential equations. Consider an electron moving in an ensemble of potentials $V_n(\boldsymbol{r} - \boldsymbol{r}_n)$. The equation for the Green function is

$$\{E + \nabla^2 - \sum_n{}' V_n(\boldsymbol{r} - \boldsymbol{r}_n) + i\varepsilon\}\, G(\boldsymbol{r}, \boldsymbol{r}') = \delta(\boldsymbol{r} - \boldsymbol{r}'). \tag{7.2.1}$$

In the same way as in the case of difference equations, we may express the spectral density $D(E)$ in terms of the Green function:

$$D(E) = \int \varrho(\boldsymbol{f}, E)\, d\boldsymbol{f}, \tag{7.2.2}$$

where

$$\varrho(\boldsymbol{f}, E) \equiv -\operatorname{Im} G(\boldsymbol{f})/\pi, \quad G(\boldsymbol{f}) \equiv \int \exp\{-i\boldsymbol{f}\cdot(\boldsymbol{r} - \boldsymbol{r}')\} \times G(\boldsymbol{r}, \boldsymbol{r}')\, d\boldsymbol{r}d\boldsymbol{r}'/V, \tag{7.2.3}$$

and V is the volume occupied by the ensemble of potentials. The function $\varrho(\boldsymbol{f}, E)$ gives the probability of finding an electron with energy E and momentum $\boldsymbol{f}$.

Following Beeby and Edwards (1963), let us introduce the T-function $T(\boldsymbol{r}, \boldsymbol{r}')$ defined by

$$G(\boldsymbol{r}, \boldsymbol{r}') = G_0(\boldsymbol{r}, \boldsymbol{r}') + \int G_0(\boldsymbol{r}, \boldsymbol{r}'')T(\boldsymbol{r}'', \boldsymbol{r}''')G_0(\boldsymbol{r}''', \boldsymbol{r}')\, d\boldsymbol{r}''d\boldsymbol{r}''', \tag{7.2.4}$$

where $G_0(\boldsymbol{r}, \boldsymbol{r}')$ is the Green function for a free electron (unperturbed system). The Fourier transform $T(\boldsymbol{f})$ of $T(\boldsymbol{r}, \boldsymbol{r}')$ is defined by the same equation as (7.2.3). In terms of Fourier transforms, the relation (7.2.4) is reduced to

$$G(\boldsymbol{f}) = G_0(\boldsymbol{f}) + G_0^2(\boldsymbol{f})T(\boldsymbol{f}), \tag{7.2.5}$$

from which it follows that

$$\operatorname{Im} T(\boldsymbol{f}) = (E - f^2)^2 \operatorname{Im} G(\boldsymbol{f}). \tag{7.2.6}$$

Thus we can use $T(f)$ instead of $G(f)$, to obtain the spectral density. The density is non-vanishing only at the poles of $T(f)$.

The development corresponding to (7.1.10) is

$$G = G_0 + G_0 \sum_n V_n G_0 + G_0 \sum_n V_n G_0 \sum_k V_k G_0$$
$$+ G_0 \sum_n V_n G_0 \sum_k V_k G_0 \sum_l V_l G_0 + \dots, \qquad (7.2.7)$$

where the symbolic notation is used, in which integrations are implicitly contained.

Let us define the T-function for a single potential $t_n(\boldsymbol{r}, \boldsymbol{r}')$ by

$$g_n(\boldsymbol{r}, \boldsymbol{r}') = G_0(\boldsymbol{r} - \boldsymbol{r}') + \int G_0(\boldsymbol{r}, \boldsymbol{r}'')\, t_n(\boldsymbol{r}'', \boldsymbol{r}''')\, G_0(\boldsymbol{r}''', \boldsymbol{r}')\, d\boldsymbol{r}'' d\boldsymbol{r}''', \qquad (7.2.8)$$

where $g_n(\boldsymbol{r}, \boldsymbol{r}')$ is the Green function for the potential V_n, i.e. the solution of the equation

$$(E + \nabla^2 - V_n(\boldsymbol{r} - \boldsymbol{r}_n) + i\varepsilon)\, g_n(\boldsymbol{r}, \boldsymbol{r}') = \delta(\boldsymbol{r} - \boldsymbol{r}'). \qquad (7.2.9)$$

Then it can be shown that (7.2.7) can be manipulated to give the expansion in terms of t-functions:

$$G = G_0 + G_0 \sum_n t_n G_0 + G_0 \sum_{n \neq k} (t_n G_0 t_k) G_0$$
$$+ G_0 \sum_{\substack{n \neq k \\ k \neq l}} (t_n G_0 t_k G_0 t_l)\, G_0 + \cdots \qquad (7.2.10)$$

From (7.2.10) and the definition of T-function it follows that

$$T = \sum_n t_n + \sum_{n \neq k} t_n G_0 t_k + \sum_{\substack{n \neq k \\ k \neq l}} t_n G_0 t_k G_0 t_l + \cdots \qquad (7.2.11)$$

As an example consider the case of a one-dimensional liquid-metal model. The equation for $g(x, x')$ is given by

$$(E + \nabla^2 + V\delta(x - a) + i\varepsilon)\, g(x, x') = \delta(x - x'). \qquad (7.2.12)$$

The relation corresponding to (7.2.8) is

$$g(x, x') = G_0(x, x') - V \int G_0(x, x'')\, \delta(x'' - a)\, g(x'', x')\, dx'', \qquad (7.2.13)$$

so that

$$g(a, x') = G_0(a, x') - V G_0(a, a)\, g(a, x'). \qquad (7.2.14)$$

Since

$$G_0(x, x') = -(i/2k)\exp\{ik|x - x'|\}, \qquad E > 0,$$
$$= -(1/2\varkappa)\exp\{-\varkappa|x - x'|\}, \quad E < 0, \quad (7.2.15)$$

$$G_0(a, a) = -i/2k, \qquad (7.2.16)$$

we have

$$g(x, x') = G_0(x, x')$$
$$-V\int \frac{G_0(x, x'')\,\delta(x'' - a)\,\delta(x''' - a)\,G_0(x''', x')}{1 - Vi/2k}\,dx''dx''', \qquad (7.2.17)$$

and consequently

$$t(x, x') = -V\delta(x - a)\,\delta(x' - a)/\{1 - Vi/2k\}. \qquad (7.2.18)$$

From (7.2.18), (7.2.11) and the definition of $T(f)$, one gets

$$T(f) = \frac{Nt}{V}\Big[1 + t\sum_{k(\neq n)} \exp\{-if(x_n - x_k)\}\,G_0(x_n - x_k)$$
$$+ t^2 \sum_{\substack{k(\neq n)\\ l(\neq k)}} \exp\{-if(x_n - x_k)\}\,G_0(x_n - x_k)$$
$$\times \exp\{-if(x_k - x_l)\}\,G_0(x_k - x_l) + \cdots, \qquad (7.2.19)$$

where t is the Fourier transform of $t(x, x')$, which is now a constant $-V/\{1 - Vi/2k\}$, and f is the scalar momentum.

It can be shown that for the case of a regular lattice, (7.2.19) becomes a geometric series, so that it can be summed up exactly. In the disordered system, however, we have to average $T(f)$ over the configurations of potentials, and the series for $T(f)$ is no longer geometric, so that it cannot be summed up. Notwithstanding this, Beeby and Edwards treated it as if it were geometric, and obtained the approximate expression

$$\langle T(f)\rangle \cong \frac{t}{1 - \dfrac{t}{N}\sum_{n\neq k} G_0(x_n - x_k)\cos f(x_n - x_k)}. \qquad (7.2.20)$$

The energy levels are at the zeros of the denominator of (7.2.20). Beeby and Edwards showed that, for the negative-energy region, the interval of $\varkappa$ in which this denominator can vanish is bounded by two curves

$$\text{(i)}\quad (2\varkappa/V) - 1 = 2\exp(-\varkappa l_{\min}),$$

$$\begin{aligned}\text{(ii)}\quad (2\varkappa/V) - 1 &= -2\exp(-\varkappa\pi/f), \qquad f \leqq \pi/l_{\min}, \\ &= -2\exp(-\varkappa l_{\min}), \qquad f \geqq \pi/l_{\min},\end{aligned} \tag{7.2.21}$$

where $l_{\min}$ is the minimum distance between potentials. If ζ is defined such that the equation

$$(2\varkappa/V) - 1 = -2\exp(-\varkappa\zeta)$$

has only one root for $\varkappa$, then equation (ii) has zero, one or two solutions according as π/f or $l_{\min}$ is less than, equal to, or greater than ζ. The boundary curves are shown in Fig. 7.4 for the cases $l_{\min} > \zeta$ and $l_{\min} < \zeta$. In the shaded region $\varrho(f, E) = 0$. It is seen that an energy gap persists until the minimum distance reaches the value ζ.

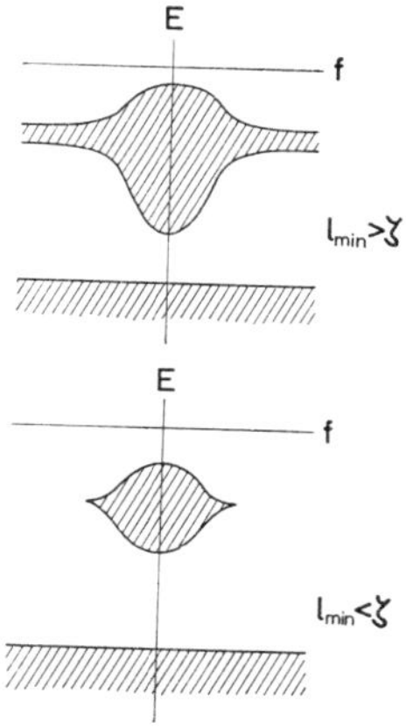

FIG. 7.4. The boundary curves of the region in which the function $\varrho(f, E)$ is non-vanishing. In the shaded region $\varrho(f, E) = 0$. (After Beeby and Edwards (1963).)

To compare the above result with that obtained in § 5.2, let us calculate the top of the zeroth gap and the bottom of the first gap from (5.2.14). For sufficiently large $\varkappa$, it becomes

$$(2\varkappa/V) - 1 \cong \pm 2\exp(-\varkappa l_{\min}), \tag{7.2.22}$$

in accordance with (7.2.21). Thus when V is so large that the whole band lies sufficiently deep in the negative-energy region, the geometric approximation gives the correct result for the bounds of the first band. On the contrary, for the lower bound of the second band, the above result is at variance with that of § 5.2, since according to the phase theory, it should be determined by the maximum spacing l_{max} rather than by l_{min}. Moreover, the lower bound of the second band should be given by $\varkappa = 0$, inasmuch as in the case treated by Beeby and Edwards l_{max} is assumed to be infinity. Thus the theory of Beeby and Edwards gives only a weaker statement about the upward extension of the first gap.

Edwards (1961) considered a weak-binding approximation and gave a suggestion that the gap is immediately destroyed upon the introduction of the slightest disorder. This conclusion appears to be conflicting with the phase theoretical one. However, the argument of Edwards is only qualitative, and the precise meaning of the word "slightest disorder" is obscure. If the distribution of spacing is not bounded, the gap may be destroyed even if the disorder is so slight that the distribution is strongly concentrated at the mean spacing, and the Edwards' suggestion may be correct.

Sah and Eisenschitz (1960) also considered a similar approximation, but they could not reach any definite conclusion concerning the gap owing to the low degree of approximation used by them. Gubanov (1954, 1961) treated the problem by the ordinary perturbation theory, and concluded that as the disorder is increased, all the energy levels are raised, its amount being, however, larger at the top than at the bottom of each band, so that the gaps shrink. This result contradicts with phase-theoretical and numerical results, which show that the top rises while the bottom falls with the increase of disorder. Thus it may be said that the approximate theories in general give results which are not only at variance with that of the phase theory and numerical calculation, but also contradict with one another.

Matsubara and Toyozawa (1961) investigated the problem of impurity band of a semiconductor by using a diagrammatical procedure similar to that described in the preceding section. Klauder (1961) also devised a diagram method, considering several different procedures of approximation, to investigate the impurity band of the model treated by Frisch and Lloyd (1960) (see § 2.4). For high concentrations of impurity, the best one of his procedures gave a result which agrees well with the numerical result of Frisch and Lloyd. The method of Matsubara and Toyozawa just corresponds to this procedure, as was shown by Yonezawa (1964). A different approximate procedure is necessary in order to obtain a good agreement for low concentrations. Such a procedure was presented by des Cloizeaux (1965) (method of self-propagator). Thus the diagram method works well for the problem in which the spectrum is expected to have only a relatively smooth structure. It should be noted, however, that it is not *a priori* clear which procedure of approximation has to be used for a given condition, e.g. for a given concentration of impurity. Therefore, even when some diagrammatic technique is discovered which yields the finer structure of the spectrum, one will not be able to put full confidence in the result obtained by it. In any case, it cannot be really expected that the Green-function or perturbation theory will provide us with reliable information concerning the detailed structure of the spectrum of a disordered system.

Also the three-dimensional electronic systems have been investigated by the method of Green function (Edwards (1962), Takeno (1962b, c), Pharizeau and Ziman (1963), Beeby (1964), and Ballentine and Heine (1964)). As for the structure of the spectrum, however, no detailed argument has been given by these authors.

Lifshitz (1942a, b, c, 1947a, b, 1948, 1952, 1956, 1963, 1964) and Lifshitz and Stepanova (1946a, b, 1957) carried out a series of very general and detailed examinations of the spectral properties of the lattices containing impurities in small concentrations. Upon the basis of more or less intuitive arguments, they obtained several interesting conclusions concerning the effect of impurities on the spectrum near the singular points such as band edges and impurity levels, at which the ordinary perturbational methods break down. As the physical picture is always clear in their arguments, the theory will be helpful for the investigation of detailed structures of the spectrum.

7.3. Method of Averaged Eigenvalue Equation

In § 1.5 it was shown that when the cyclic boundary condition is used, the eigenvalues of the system are given by the roots of the equation

$$\text{trace } \mathbf{T}_{N;1} = 2\mathbf{I}. \tag{7.3.1}$$

In the disordered system, the left-hand side of this equation is a function of the configuration in each individual system. If we take its average over all possible configurations, we obtain

$$\langle f \rangle = \langle \text{trace } \mathbf{T}_{N;1} \rangle = 2, \tag{7.3.2}$$

where for simplicity the one-dimensional system is assumed. The roots of (7.3.2) do not give, of course, a correct averaged spectrum, inasmuch as the average over the distribution of roots of an individual equation is not equal to the distribution of roots of an averaged equation. Indeed, it was pointed out by Dean (1963) that the correct method of averaging the eigenvalue equation is not to take the arithmetic mean but to take the geometric one. It may still be considered, however, to give an approximate averaged spectrum, though the nature of approximation is not clear.

As an example, consider the isotopically disordered diatomic chain composed of atoms with masses $m^{(0)}$ and $m^{(1)}$, and use the representation (1.3.11), denoting the transfer matrices associated with the light and heavy atoms by $\mathbf{T}^{(0)}$ and $\mathbf{T}^{(1)}$ respectively. Let the total number of light atoms be r. Then there are $\binom{N}{r}$ configurations that the chain can possess, the corresponding arrays of transfer matrices being given by the coefficient of z^r in the expansion of $(\mathbf{T}^{(1)} + z\mathbf{T}^{(0)})^N$. The function $\langle f \rangle$ is thus given by

$$\langle f \rangle = \text{trace} \frac{1}{\binom{N}{r}(2\pi i)} \oint \frac{(\mathbf{T}^{(1)} + z\mathbf{T}^{(0)})^N}{z^{r+1}} \, dz, \tag{7.3.3}$$

where the contour of integration circles the origin once in a counterclockwise sense.

Denote the determinant of $\mathbf{T}^{(1)} + z\mathbf{T}^{(0)}$ by D^2 (which comes out to be $(1 + z)^2$) and introduce a matrix $\mathbf{F}(z)$ defined by

$$\mathbf{F}(z) \equiv (\mathbf{T}^{(1)} + z\mathbf{T}^{(0)} - D\mathbf{I} \cos p)/iD \sin p, \tag{7.3.4}$$

where

$$\cos p \equiv (g + zh)/D = (g + zh)/(1 + z), \tag{7.3.5}$$

and

$$g \equiv \tfrac{1}{2} \text{trace } \mathbf{T}^{(1)} = 1 - (m^{(1)}\omega^2)/2K,$$

$$h \equiv \tfrac{1}{2} \text{trace } \mathbf{T}^{(0)} = 1 - (m^{(0)}\omega^2)/2K. \tag{7.3.6}$$

The function $p(z)$ can be made single-valued by appropriately choosing the Riemann sheets. The matrix $\mathbf{F}(z)$ has the properties

$$\text{trace } \mathbf{F}(z) = 0 \tag{7.3.7}$$

and

$$\mathbf{F}^2(z) = \mathbf{I} \tag{7.3.8}$$

for all z. By repeated use of (7.3.8), it can be shown that

$$(\mathbf{T}^{(1)} + z\mathbf{T}^{(0)})^N = D^N\{\mathbf{I}\cos Np + i\mathbf{F}(z)\sin Np\}, \tag{7.3.9}$$

so that combining (7.3.3) and (7.3.9) and using (7.3.7), we obtain the result

$$\langle f \rangle = \frac{1}{\binom{N}{r}(\pi i)} \oint \frac{D^N \cos Np}{z^{r+1}}\, dz = K_+ + K_-, \tag{7.3.10}$$

where

$$K_\pm \equiv \frac{1}{\binom{N}{r}(2\pi i)} \int \frac{\exp\{Nh_\pm(z)\}}{z}\, dz, \tag{7.3.11}$$

with

$$h_\pm(z) \equiv \pm ip + \ln D - (r+1)\ln z/N. \tag{7.3.12}$$

Flinn *et al.* (1961) and Schlup (1965) evaluated the integrals $K_\pm$ in the limit of large N by applying the saddle-point method and then calculated the spectra for various values of $Q^{(1)}$ and $f_L = r/N$. Their result for $Q^{(1)} = 2$ and $f_L = 0{\cdot}1$ is shown in Fig. 7.5, which

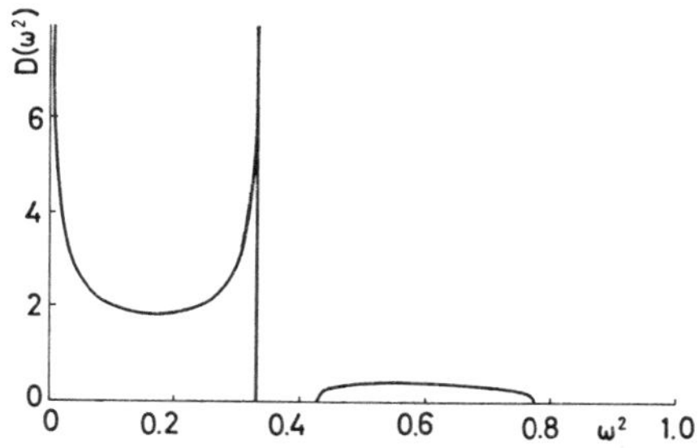

FIG. 7.5. The density of eigenfrequency spectrum of the isotopically disordered chain obtained by the method of averaged eigenvalue equation for the case $Q^{(1)} = 2$ and $f_L = 0{\cdot}1$. (After Flinn *et al.* (1961).)

is to be compared with Fig. 7.3 and Fig. 2.6a. It is seen that the method of averaged eigenvalue equation can give the spectrum with an accuracy at most corresponding to that of the self-consistent field approximation of Davies and Langer.

Faulkner and Korringa (1961) calculated the spectral gaps of a disordered diatomic Kronig–Penney model by the same method.

They obtained for the spectral gap of a disordered system the result illustrated in Fig. 7.6. It is seen from this figure that the gaps are slightly wider than those predicted by the Saxon–Hutner theorem. Such a situation is more conspicuous in the case of the liquid-metal model, as has been shown by Hiroike (1965). He calculated the gaps of liquid metals in which the distribution of spacing is Gaussian with the dispersion σ^2. The results for $V\langle l_n\rangle = 0{\cdot}25$ and 4, for the first and second gaps are illustrated in Fig. 7.7,

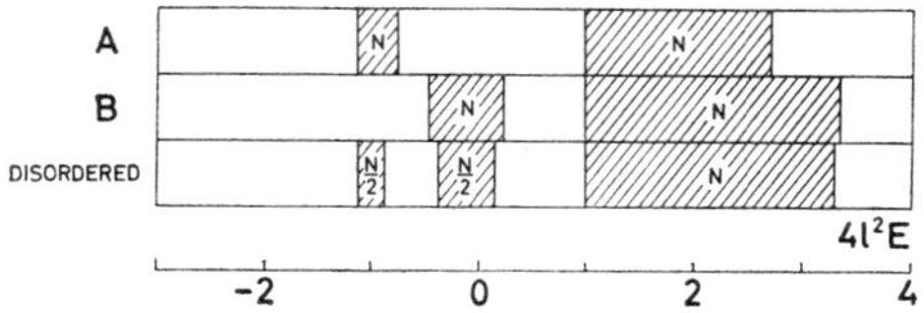

FIG. 7.6. Spectral bands and gaps obtained by the method of averaged eigenvalue equation for the disordered diatomic Kronig–Penney model with $lV^{(A)} = 2\pi$, $lV^{(B)} = \pi$ and $f_A = 0{\cdot}5$. The number in each band gives the number of states. (After Faulkner and Korringa (1961).)

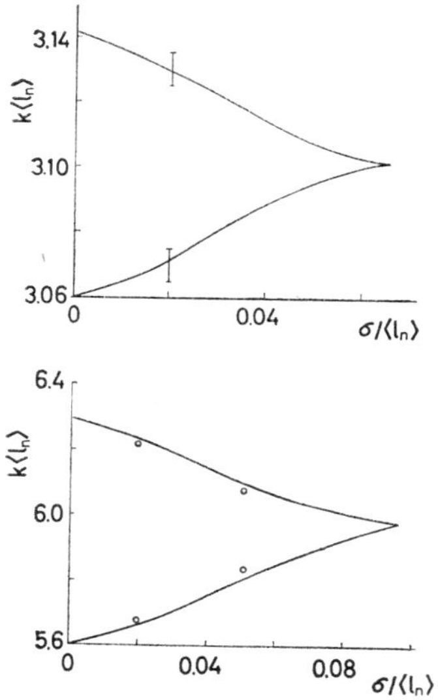

FIG. 7.7. The variation with σ of the boundaries of (a) the first gap for $V\langle l_n\rangle = 0{\cdot}25$ and (b) the second gap for $V\langle l_n\rangle = 4$. The bars in (a) and the circles in (b) represent the results obtained by Makinson and Roberts (1960) by numerical computation. (After Hiroike (1965).)

together with those of Makinson and Roberts already mentioned in § 2.5. It is seen that these are remarkably in agreement with each other. Such an agreement suggests that, for the problem of spectral gaps, the methods of averaged eigenvalue equation can give correct results. It cannot, however, provide us with valuable informations about the detailed behaviours of spectral density inside the band, as is shown by the above-mentioned result obtained by Flinn *et al.*

Faulkner (1964) also calculated the spectrum of the model considered by Lax and Phillips and Frisch and Lloyd (§ 2.5). His result is in agreement with the calculation of Frisch and Lloyd only at high concentrations of impurity, showing again that the reliability of the present method at most ranks with that of the Green-function method.

7.4. Method of Moments

As the eigenvalues of the dynamical matrix $\mathbf{P}$ are simply those of the system in question, it is clear that

$$(1/\mathrm{N})\ \mathrm{trace}\ \mathbf{P}^k = (1/\mathrm{N}) \sum_{\mu} \lambda_{\mu}^k. \tag{7.4.1}$$

The right-hand side of this equation is the kth moment of the eigenvalue distribution:

$$(1/\mathrm{N}) \sum_{\mu} \lambda_{\mu}^k \equiv \mu_k = \int_{-\infty}^{+\infty} \lambda^k D(\lambda)\, d\lambda. \tag{7.4.2}$$

Thus every moment of the spectrum can be calculated once the dynamical matrix is known.

For example, for a one-dimensional isotopic chain, we have, from (1.1.29),

$$\mu_1 = \frac{2K}{N} \sum_n \frac{1}{m_n} = 2K\left\langle \frac{1}{m_n} \right\rangle,$$

$$\mu_2 = \frac{K^2}{N}\left\{2 \sum_n \frac{1}{m_n m_{n+1}} + 4 \sum_n \frac{1}{m_n^2}\right\}$$

$$= 2K^2\left\langle \frac{1}{m_n m_{n+1}} \right\rangle + 4K^2\left\langle \frac{1}{m_n^2} \right\rangle,$$

$$\mu_3 = \frac{K^3}{N}\left\{12 \sum_n \frac{1}{m_n m_{n+1}^2} + 8 \sum_n \frac{1}{m_n^3}\right.$$

$$= 12K^3\left\langle \frac{1}{m_n m_{n+1}^2} \right\rangle + 8K^3\left\langle \frac{1}{m_n^3} \right\rangle, \quad \text{etc.} \tag{7.4.3}$$

When two kinds of atoms with masses $m^{(0)}$ and $m^{(1)}$ are arranged randomly with probabilities f_L and $1 - f_L$ respectively, (7.4.3) reduces to

$$\mu_1 = 2K\left\{\frac{f_L}{m^{(0)}} + \frac{(1-f_L)}{m^{(1)}}\right\},$$

$$\mu_2 = 2K^2\left\{\frac{f_L(1+2f_L)}{m^{(0)^2}} + \frac{2f_L(1-f_L)}{m^{(0)}m^{(1)}} + \frac{2(1-f_L)+(1-f_L)^2}{m^{(1)^2}}\right\},$$

$$\mu_3 = 4K^3\left\{\frac{f_L(2f_L+2)}{m^{(0)^3}} + \frac{3(1-f_L)^2+2(1-f_L)}{m^{(1)^3}} + \frac{3f_L(1-f_L)}{m^{(0)}m^{(1)^2}} + \frac{3f_L(1-f_L)}{m^{(0)^2}m^{(1)}}\right\}, \quad \text{etc.} \tag{7.4.4}$$

in the limit $N \to \infty$.

There are several methods for calculating an approximate spectrum from these moments. Here we explain only one of them called *polynomial approximation* for the case of the vibrational spectrum $D(\omega)$. As this is an even function and extends between $-\omega_{\max}$ and $+\omega_{\max}$, all the odd moments vanish, while the even ones are given by

$$\mu_k = \frac{\int_{-\omega_{\max}}^{\omega_{\max}} \omega^{2k} D(\omega)\, d\omega}{\int_{-\omega_{\max}}^{\omega_{\max}} D(\omega)\, d\omega} = \int_0^{\omega_{\max}} \omega^{2k} D(\omega)\, d\omega. \tag{7.4.5}$$

If we expand $D(\omega)$ in terms of Legendre polynomials as

$$D(\omega) = \sum_{n=0}^{\infty} a_n P_n(\omega/\omega_{\max}), \tag{7.4.6}$$

the coefficients are given by

$$a_{2n+1} = 0,$$

$$a_{2n} = \frac{4n+1}{2} \int_{-1}^{+1} D(x\omega_{\max})\, P_{2n}(x)\, dx. \tag{7.4.7}$$

As $P_{2n}(x)$ is a polynomial of order $2n$, a_{2n} is determined by the moments $\mu_0, \mu_1, \ldots, \mu_n$, and does not depend on the higher moments. Thus if one omits the terms higher than the $2m$th in the expansion (7.4.6), the approximate spectrum can be obtained by calculating only the moments of up to the *m*th. One can also use the other polynomials such as the Hermite polynomial.

Domb *et al.* (1959a, b) calculated the spectra of isotopically disordered diatomic lattices by using the polynomial method, together with some subsidiary techniques, and by computing first twenty moments. Their result for $Q^{(1)} = 2$ and $f_L = 0{\cdot}1$ is shown in Fig. 7.8. Comparison with Figs. 7.3 and 2.6a shows that the present method can elucidate a gross feature of the peak structure. The rightmost peak seemingly corresponds to the high-frequency peaks in Dean's spectrum, though its position is inaccurate. The minor peaks which exist in Dean's spectrum just outside the main band now seem to be included in the central broad peak. On the whole, however, the effectiveness of the method of moments is at most comparable with that of the Green-function method.

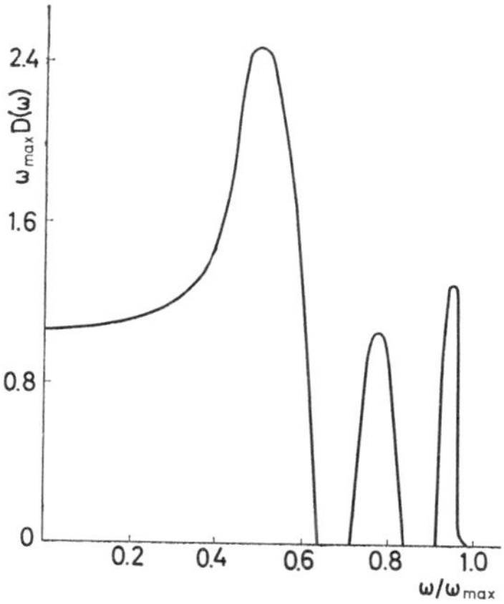

FIG. 7.8. The density of eigenfrequency spectrum of the isotopically disordered diatomic chain for the case $Q^{(1)} = 2$ and $f_L = 0{\cdot}1$, obtained by the method of moments. (After Domb *et al.* (1959a).)

Fukuda and Yoshida (1962), Fukuda (1962b), and Hori (1960) tried to improve the method by an artifice of applying it only for high-frequency region, approximating the low-frequency part by the spectrum of a virtual regular lattice composed of atoms with average mass. They failed, however, to obtain any significantly better result. Bradley (1961) applied the method of moments to a disordered three-dimensional simple cubic lattice, with a result which likewise lacks any detailed structure.

Ford (1959) treated the problem of the electronic spectrum of a liquid-like array of N identical square wells by the moment method, using the tight-binding approximation mentioned in § 1.2. He took as the atomic orbital the first excited state of the square-well potential, and assumed that the distance between potentials is distributed according to the Gaussian distribution with the standard deviation σ. From the Hamiltonian matrix obtained, he calculated first six moments and constructed the spectrum by polynomial methods. Figures 7.9 and 7.10 show some of his results compared with the numerical calculation made by Landauer and Helland mentioned in § 2.5. It seems that the method of moments can give a fairly accurate spectrum in the liquid case

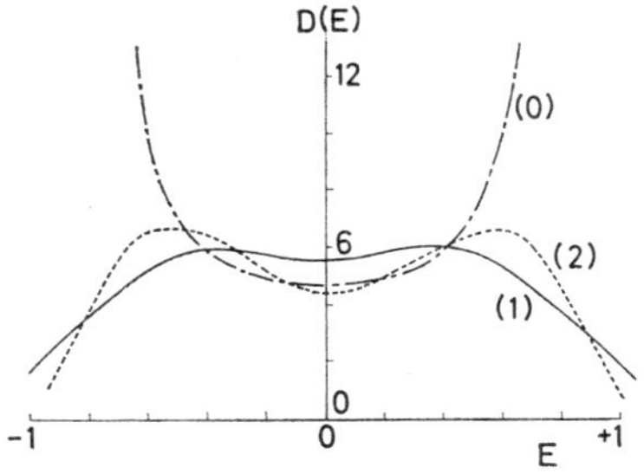

FIG. 7.9. The curves (1) and (2) are the densities of states obtained by the moment method using the Hermite and Legendre expansions, respectively, for the liquid-like array of square-well potentials with $\sigma = 0{\cdot}471$. The curve (0) is the density for $\sigma = 0$. (After Ford (1959).)

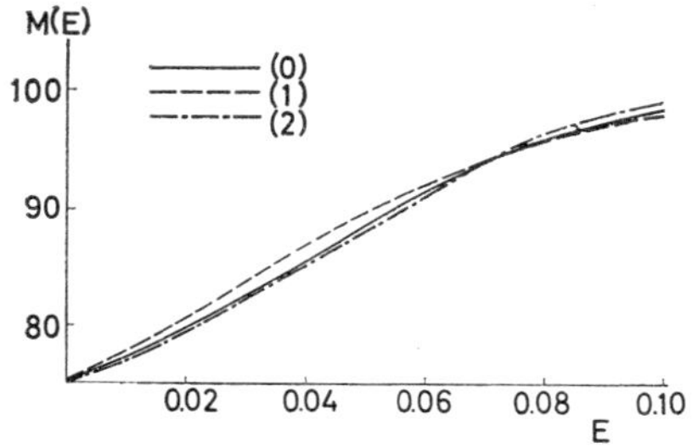

FIG. 7.10. The curves (1) and (2) are the integrated densities obtained by the moment method using the Hermite and Legendre expansions respectively, for the liquid-like array of square-well potentials with $\sigma = 0{\cdot}471$. The curve (0) is the result of the numerical calculation of Landauer and Helland (1954). (After Ford (1959).)

where the spectrum is more or less smooth and has no fine structures.

Since the averaged eigenvalue equation (7.3.2) is a polynomial equation of order N with the variable ω^2, it is clear that the moments of squared-eigenfrequency distribution $D(\omega^2)$ can be expressed in terms of its coefficients. By calculating the coefficients for the isotopically disordered diatomic chain, Mahanty (1960) obtained an expression for the moments. Such an expression is, of course, only an approximate one, giving the moments of the spectrum which are to be obtained by the method of averaged eigenvalue equation. As was pointed out by Flinn *et al.* (1961) and Schlup (1963a), however, the first few moments coincide, notwithstanding, with the exact ones.

Schlup (1963a) made a detailed examination of the moment method for the case of isotopically disordered polyatomic chains, giving the expressions of moments in terms of the moments, central moments and semi-invariants of the reciprocal mass. Pirenne (1958,) Litzman and Klvana (1962), and Domb (1963) also made detailed investigations on the expansion of the spectrum in terms of moments and the method of calculation of moments. These arguments are, however, purely formal, without any actual calculations.

7.5. Theories in Effective-mass Approximation

In connection with the problem of impurity bands in the effective-mass approximation, still other approximate theories have been presented. Starting from his functional equation, Schmidt (1957) obtained an approximate expression for the integrated density in the limit of low density of impurities:

$$\begin{aligned} M(\varkappa/\varkappa_0) &= \{1 - |(\varkappa/\varkappa_0 - 1)|^{c/\varkappa_0}\}/\{\tfrac{3}{2} - \tfrac{1}{2}|(\varkappa/\varkappa_0 - 1)|^{c/\varkappa_0}\}^2, \\ &\qquad 1 \leqq \varkappa/\varkappa_0 \leqq 2, \\ &= 1/\{\tfrac{3}{2} - \tfrac{1}{2}|(\varkappa/\varkappa_0 - 1)|^{c/\varkappa_0}\}^2, \\ &\qquad 0 \leqq \varkappa/\varkappa_0 \leqq 1, \end{aligned} \tag{7.5.1}$$

where c is the concentration of the impurity potentials and $\varkappa_0 \equiv (-E_0)^{1/2}$, E_0 being the energy of the impurity level given by $-V^2/4$. Morrison (1962) obtained the same formula by starting from the diffusion equation of Frisch and Lloyd (1960) mentioned in § 2.5.

Lax and Phillips (1958) treated the problem in multiple-scattering formalism. The Schrödinger equation for the field of δ-potentials situated at random positions

$$\left\{\nabla^2 + 2\varkappa_0 \sum_n \delta(x - x_n)\right\} \psi(x) = \varkappa^2 \psi(x) \tag{7.5.2}$$

can be written in the form

$$\psi(x) = \varphi(x) + (\varkappa_0/\varkappa) \int \exp(-\varkappa|x - x'|) \sum_n \delta(x' - x_n)\, \psi(x')\, dx', \tag{7.5.3}$$

by using the Green function (7.2.15). Here $\varphi(x)$ is the external incident wave (if any). By putting

$$L_n(x) \equiv (\varkappa_0/\varkappa) \exp(-\varkappa\,|x - x_n|)\, \psi(x_n) = \exp(-\varkappa\,|x - x_n|)\, L_n(x_n), \tag{7.5.4}$$

equation (7.5.3) may be written

$$\psi(x) = \varphi(x) + \sum_n L_n(x). \tag{7.5.5}$$

Now define the effective field ψ^n on the nth scatterer by

$$\psi^n(x) \equiv \psi(x) - L_n(x) = \varphi(x) + \sum_{m \neq n} L_m(x). \tag{7.5.6}$$

Then from (7.5.5) we find

$$L_n(x) = (\varkappa/\varkappa_0 - 1)^{-1} \exp(-\varkappa\,|x - x_n|)\, \psi^n(x_n) \equiv A(|x - x_n|)\, \psi^n(x_n). \tag{7.5.7}$$

If we put $x = x_n$ in (7.5.5) and use (7.5.7), we obtain an equation for the effective fields:

$$\psi^n(x_n) = \varphi(x_n) + \sum_{n \neq m} A_{nm} \psi^m(x_m), \tag{7.5.8}$$

where $A_{nm} \equiv A(|x_n - x_m|)$. The choice $\varphi(x) = 0$, which corresponds to the problem of eigenfunctions, leads to the equation

$$c_n = \sum_{m=1}^{N} A_{nm} c_m, \tag{7.5.9}$$

where $c_n \equiv \psi^n(x_n)$.

For low densities, the broadening of the bound state of isolated potential will be produced mainly by pair interactions. If the mth and nth potentials form a pair, we may expect that $c_m, c_n \gg c_k$ $(k \neq m, n)$ for solutions localized around them. Thus $c_m = A_{mn}c_n$ and $c_n = A_{nm}c_m$, so that

$$\varkappa/\varkappa_0 - 1 = \pm \exp(-\varkappa |x_m - x_n|). \qquad (7.5.10)$$

The probability that the nearest neighbour of an atom located at the origin lies between x and $x + dx$ is given by $2c \exp(-2c|x|)$, and the probability that the nearest-neighbour distance is less than x is $\exp(-2cx)$. Therefore, the fraction of states for which $\varkappa/\varkappa_0$ is larger than the indicated value is given by

$$\begin{aligned} M(\varkappa/\varkappa_0) &= \tfrac{1}{2} - \tfrac{1}{2}|\varkappa/\varkappa_0 - 1|^{2c/\varkappa}, \quad 1 \leqq \varkappa/\varkappa_0 \leqq 2, \\ &= \tfrac{1}{2} + \tfrac{1}{2}|\varkappa/\varkappa_0 - 1|^{2c/\varkappa}, \quad 0 \leqq \varkappa/\varkappa_0 \leqq 1. \qquad (7.5.11) \end{aligned}$$

In Fig. 7.11, the curves obtained from (7.5.1) and (7.5.11) are compared with the result of numerical calculation already mentioned in § 2.5 (Fig. 2.17). It is seen that the pair approximation is good only for the wings of the impurity band, while Schmidt's solution agrees well with the numerical calculation everywhere.

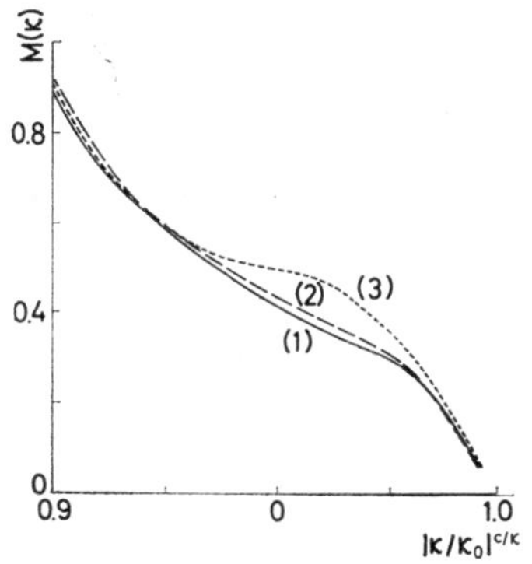

FIG. 7.11. Integrated density of states for the one-dimensional random array of δ-potentials obtained (1) by machine calculation, (2) from the formula of Schmidt, and (3) from the pair approximation. The concentration c of potentials is taken to be $0{\cdot}1\varkappa_0$. (After Lax and Phillips (1958).)

At high densities the fractional fluctuation of the number of atoms in a given spatial range is small, so that one can put $|c_n| = C$ and $c_n = C \exp(ik'x_n)$. Then one obtains

$$\varkappa/\varkappa_0 - 1 = -1 + \sum_{n,m} \exp\{-\varkappa|x_m - x_n| + ik'(x_m - x_n)\}. \qquad (7.5.12)$$

The sum may be approximated by its ensemble average:

$$\varkappa/\varkappa_0 - 1 = -1 + \int_{-\infty}^{\infty} c \exp(-\varkappa |x| + ik'x)\, dx, \qquad (7.5.13)$$

so that

$$\varkappa^2 + k'^2 = 2c\varkappa_0. \qquad (7.5.14)$$

Thus the numbers of zeros in $\cos k'x$ per impurity atom is given by

$$M(\varkappa/\varkappa_0) = k'/\pi n = (2c\varkappa_0 - \varkappa^2)^{1/2}/\pi n. \qquad (7.5.15)$$

The result of this *optical-model approximation* has already been plotted in Fig. 2.17. It is seen that this approximation is good for high densities except at the band edge. Lax and Phillips improved the approximation by taking into account the fluctuation, and obtained a formula which reproduces fairly well the exact curve.

It may be concluded therefore that for the problem of impurity band in the effective-mass or similar approximations, in which we need not bother about the fine structures of the spectrum and other subtle problems, various approximate theories including Green-function theory may work well in the respective domains of applicability.

Bibliographical Notes

For the materials which lie outside the scope of the present book there exists a vast number of papers and books. Here we mention only the following representative works:

[1] Maradudin, A. A., Montroll, E. W. and Weiss, G. H. (1963) *Theory of Lattice Dynamics in the Harmonic Approximation* (vol. 3 of the Supplement of *Solid State Physics*, edited by F. Seitz and D. Turnbull), Academic Press, New York.

[2] Maradudin, A. A. (1963) *Topics in the Theory of the Vibrations of Imperfect Crystals, Astrophysics and the Manybody Problem* (vol. 2 of the *Brandeis University Summer Institute Lectures in Theoretical Physics*, edited by K. W. Ford), W. A. Benjamin, Inc., New York, 107.

[3] Wallis, R. F. (Editor) (1965) Lattice dynamics, *Proceedings of the International Conference held at Copenhagen*, 1963, Pergamon Press, Oxford.

[4] Ludwig, W. (1964) Zur Dynamik von Kristallen mit Punktdefekten, *Ergebnisse der exakten Naturwissenschaften*, Springer-Verlag, Berlin, **35**, 1.

[5] Bak, Th. A. (Editor) (1964) *Phonons and Phonon Interactions*, W. A. Benjamin, Inc., New York.

[6] Maradudin, A. A. (1966) *Theoretical and Experimental Aspects of the Effects of Point Defects and Disorder on the Vibrations of Crystals*. I. *Solid State Physics* (edited by F. Seitz and D. Turnbull), Academic Press, New York, **18**, 278.

[7] Maradudin, A. A. (1965) Some effects of point defects on the vibrations of crystal lattices, *Reports on Progress in Physics*, the Institute of Physics and the Physical Society, London, **28**, 331.

Chapter 1

The eigenfrequency spectrum of regular lattices has long been one of the central problems of lattice dynamics, so that we have a comprehensive number of textbooks and review articles. For the purpose of the better understanding of § 1.6, however, it is sufficient to read the first three chapters of [1]. A more detailed account concerning the nature of the spectrum is given in the articles by E. W. Montroll (1955), referred to in the text.

Bibliographical Notes

CHAPTERS 4 and 6

The problem of impurity modes associated with a few isolated impurities is ordinarily treated by the method of Green function. Its account is given in [1], [2], [4], [7], and the article by A. A. Maradudin in [5]. Some topics concerning the same subject are found also in [3]. It is to be mentioned that an entirely different approach has been made by M. TODA, T. KOTERA, and Y. KOGURE (1962) (*J. Phys. Soc. Japan* **17**, 426) and T. KOTERA (1962) (*Prog. Theor. Phys. Suppl.* **23**, 141), upon the basis of the analogy between the theories of lattice vibration and random walk.

CHAPTER 6

The extended impurities can be handled as well by the Green-function method as by the phase theory. The most complete treatment was given by B. Lengeler and W. Ludwig in their paper referred to in the text, which contains a list of further literature. Their work is also reported in [3].

CHAPTER 7

Inasmuch as we have at present only a few exact treatments of multi-dimensional lattices, the approximate calculation of their spectra is still important. In particular, for small concentrations of impurities, it has been proved that the approximate results can be compared with experiment with satisfactory agreements. The recent developments of approximate theories and experimental investigations are summarized in [6].

References

AGACY, R. L. and BORLAND, R. E. (1964) *Proc. Phys. Soc.* **84**, 1017.
ASAHI, T. (1962) *Prog. Theor. Phys. Suppl.* **23**, 59.
ASAHI, T. and HORI, J. (1963) Lattice dynamics, *Proceedings of the International Conference held at Copenhagen*, Pergamon Press, Oxford, 571.
ASAHI, T. (1964) *J. Res. Inst. Catalysis, Hokkaido Univ.* **11**, 133.
ASAHI, T. (1966) *Prog. Theor. Phys. Suppl.* **36**, 55.
ATKINSON, F. V. (1963) *Discrete and Continuous Boundary Problems*, Academic Press, New York.
BACON, M. D. and DEAN, P. (1962) *Nature* **194**, 541.
BACON, M. D., DEAN, P. and MARTIN, J. L. (1962) *Proc. Phys. Soc.* **80**, 174.
BALLENTINE, L. E. and HEINE, V. (1964) *Phil. Mag.* **9**, 617.
BEEBY, J. L. and EDWARDS, S. F. (1963) *Proc. Roy. Soc.* **274**, 395.
BEEBY, J. L. (1964) *Proc. Roy. Soc.* **279**, 92.
BELLMAN, R. (1956) *Phys. Rev.* **101**, 19.
BORLAND, R. E. (1961a) *Proc. Phys. Soc.* **77**, 705.
BORLAND, R. E. (1961b) *Proc. Phys. Soc.* **78**, 926.
BORLAND, R. E. (1963) *Proc. Roy. Soc.* **274**, 529.
BORLAND, R. E. (1964) *Proc. Phys. Soc.* **83**, 1027.
BORLAND, R. E. and BIRD, N. F. (1964) *Proc. Phys. Soc.* **83**, 23.
BRADLEY, J. C. (1961) *Annals of Phys.* **15**, 411.
CLOIZEAUX, J. DES (1957) *J. Phys. et Rad.* **18**, 131.
CLOIZEAUX, J. DES (1965) *Phys. Rev.* **139**, A 1531.
COURANT, R. and HILBERT, D. (1931) *Methoden der Mathematischen Physik*, vol. 1, Springer-Verlag, Berlin, 26.
DAVIES, R. W. and LANGER, J. S. (1963) *Phys. Rev.* **131**, 163.
DEAN, P. (1956) *Proc. Cambridge Phil. Soc.* **52**, 752.
DEAN, P. (1959) *Proc. Phys. Soc.* **73**, 413.
DEAN, P. (1960) *Proc. Roy. Soc.* **254**, 507.
DEAN, P. and MARTIN, J. L. (1960a) *Proc. Phys. Soc.* **75**, 452.
DEAN, P. and MARTIN, J. L. (1960b) *Proc. Roy. Soc.* **259**, 409.
DEAN, P. (1961) *Proc. Roy. Soc.* **260**, 263.
DEAN, P. (1963) Lattice dynamics, *Proceedings of the International Conference held at Copenhagen*, Pergamon Press, Oxford, 561.
DEAN, P. and BACON, M. D. (1963) *Proc. Phys. Soc.* **81**, 642.
DEAN, P. (1964) *Proc. Phys. Soc.* **84**, 727.
DEAN, P. and BACON, M. D. (1965) *Proc. Roy. Soc.* **283**, 64.
DOMB, C., MARADUDIN, A. A., MONTROLL, E. W. and WEISS, G. H. (1959a) *Phys. Rev.* **115**, 18.

References

Domb, C., Maradudin, A. A., Montroll, E. W. and Weiss. G. H. (1959b) *J. Phys. Chem. Solids* **8**, 419.

Domb, C. (1963) *Proc. Roy. Soc.* **276**, 418.

Dworin, L. (1965) *Phys. Rev.* **138**, A 1121.

Dycker, E. de and Phariseau, P. (1965) *Physica* **31**, 1337.

Dyson, F. J. (1953) *Phys. Rev.* **92**, 1331.

Edwards, S. F. (1961) *Phil. Mag.* **6**, 617.

Edwards, S. F. (1962) *Proc. Roy. Soc.* **267**, 518.

Eisenschitz, R. and Dean, P. (1957) *Proc. Phys. Soc.* **70**, 713.

Englman, R. (1958) *Nuovo Cim.* **10**, 615.

Faulkner, J. S. and Korringa, J. (1961) *Phys. Rev.* **122**, 390.

Faulkner, J. S. (1964) *Phys. Rev.* **135**, A 124.

Flinn, P. A., Maradudin, A. A. and Weiss, G. H. (1961) Scientific Paper 029-G000-P10, Westinghouse Research Laboratories, Pittsburgh.

Ford, J. (1959) *J. Chem. Phys.* **30**, 1546.

Frisch, H. L. and Lloyd, S. P. (1960) *Phys. Rev.* **120**, 1175.

Fukuda, Y. (1962a) *J. Phys. Soc. Japan* **17**, 766.

Fukuda, Y. (1962b) *Prog. Theor. Phys. Suppl.* **23**, 79.

Fukuda, Y. and Yoshida, K. (1962) *J. Phys. Soc. Japan* **17**, 920.

Fukushima, M. (1965) *Prog. Theor. Phys.* **33**, 624.

Gubanov, A. I. (1954) *J. Exp. Theor. Phys.* **26**, 139.

Gubanov, A. I. (1961) *Fiz. Tverd. Tela* **3**, 2154.

Hatakeyama, K. (1965) *J. of Hokkaido Gakugei University*, Section 2a, 16, 28.

Herzenberg, A. and Modinos, A. (1964) *Biopolymers* **2**, 561.

Hiroike, K. (1965) *Phys. Rev.* **138**, A 422.

Hori, J. (1957) *Prog. Theor. Phys.* **18**, 367.

Hori, J. and Asahi, T. (1957) *Prog. Theor. Phys.* **17**, 523.

Hori, J. (1960) *Prog. Theor. Phys.* **23**, 475.

Hori, J. (1961) *J. Phys. Soc. Japan* **16**, 23.

Hori, J. (1962) *Prog. Theor. Phys. Suppl.* **23**, 3.

Hori, J. and Fukushima, M. (1963) *J. Phys. Soc. Japan* **19**, 296.

Hori, J. (1964a) *Prog. Theor. Phys.* **32**, 371.

Hori, J. (1964b) *Prog. Theor. Phys.* **32**, 471.

Hori, J. and Asahi, T. (1964) *Prog. Theor. Phys.* **31**, 49.

Hori, J. and Matsuda, H. (1964) *Prog. Theor. Phys.* **32**, 183.

Hori, J. (1966) *Prog. Theor. Phys. Suppl.* **36**, 3.

Hori, J. and Fukushima, M. (1966) *JAERI 1113*, Japan Atomic Energy Research Institute, Tokai-mura, 55.

James, H. M. and Ginzbarg, A. S. (1953) *J. Phys. Chem.* **57**, 840.

Kerner, E. H. (1954) *Phys. Rev.* **95**, 687.

Kerner, E. H. (1956) *Proc. Phys. Soc.* **69**, 234.

Klauder, J. R. (1961) *Annals of Phys.* **14**, 43.

Kobori, I. (1965) *Prog. Theor. Phys.* **33**, 614.

Kuliasko, F. (1964a) *Physica* **30**, 2180.

Kuliasko, F. (1964b) *Physica* **30**, 2185.

Landauer, R. and Helland, J. C. (1954) *J. Chem. Phys.* **22**, 1655.

Langer, J. S. (1961) *J. Math. Phys.* **2**, 584.

Lax, M. and Phillips, J. C. (1958) *Phys. Rev.* **110**, 41.

LENGELER, B. and LUDWIG, W. (1964) *Phys. stat. Solidi* **7**, 463.
LIFSHITZ, I. M. (1942a) *J. Exp. Theor. Phys.* **12**, 117.
LIFSHITZ, I. M. (1942b) *J. Exp. Theor. Phys.* **12**, 137.
LIFSHITZ, I. M. (1942c) *J. Exp. Theor. Phys.* **12**, 156.
LIFSHITZ, I. M. (1947a) *J. Exp. Theor. Phys.* **17**, 1017.
LIFSHITZ, I. M. (1947b) *J. Exp. Theor. Phys.* **17**, 1076.
LIFSHITZ, I. M. (1948) *J. Exp. Theor. Phys.* **18**, 293
LIFSHITZ, I. M. (1952) *Uspekhi Mat. Nauk* **7**, 170.
LIFSHITZ, I. M. (1956) *Nuovo Cim. Suppl.* **3**, 716.
LIFSHITZ, I. M. and STEPANOVA, G. I. (1956a) *J. Exp. Theor. Phys.* **30**, 938.
LIFSHITZ, I. M. and STEPANOVA, G. I. (1956b) *J. Exp. Theor. Phys.* **31**, 156.
LIFSHITZ, I. M. (1963) *J. Exp. Theor. Phys.* **44**, 1723.
LIFSHITZ, I. M. (1964) *Advances in Phys.* **13**, 483.
LITZMANN, O. and KLVANA, F. (1962) *Physica stat. Solidi* **2**, 42.
LUTTINGER, J. M. (1951) *Philips Res. Rep.* **6**, 303.
MAHANTY, J. (1961) *Nuovo Cim.* **19**, 46.
MAKINSON, R. E. B. and ROBERTS, A. P. (1960) *Australian J. Phys.* **13**, 437.
MARADUDIN, A. A. and WEISS, G. H. (1958) *J. Chem. Phys.* **29**, 631.
MARTIN, J. L. (1961) *Proc. Roy. Soc.* **260**, 139.
MATSUBARA, T. and TOYOZAWA, Y. (1961) *Prog. Theor. Phys.* **26**, 739.
MATSUDA, H. (1962a) *Prog. Theor. Phys.* **27**, 811.
MATSUDA, H. (1962b) *Prog. Theor. Phys. Suppl.* **23**, 22.
MATSUDA, H. (1964) *Prog. Theor. Phys.* **31**, 161
MATSUDA, H. (1965) Unpublished.
MATSUDA, H. and OKADA, K. (1965) *Prog. Theor. Phys.* **34**, 539.
MATSUDA, H. and TERAMOTO, E. (1965) *Prog. Theor. Phys.* **34**, 314.
MATSUDA, H. (1966) *Prog. Theor. Phys.* **36**, 1070.
MONTROLL, E. W. (1955) *Proceedings of the Third Berkeley Symposium on Mathematical Statistics and Probability*, University of California Press, Berkeley, 209.
MONTROLL, E. W. and POTTS, R. B. (1955) *Phys. Rev.* **100**, 525.
MORRISON, J. A. (1962) *J. Math. Phys.* **3**, 1023.
MORSE, P. M. and FESHBACH, H. (1953) *Methods of Theoretical Physics*, McGraw-Hill Book Co., New York, Chapter 6.
MOTT, N. F. and TWOSE, W. D. (1961) *Advances in Phys.* **10**, 107.
MOZER, B. and MARADUDIN, A. A. (1963) *Bull. Am. Phys. Soc.* [II] **8**, 193.
OSAWA, K. and KOTERA, T. (1966) *Prog. Theor. Phys. Suppl.* **36**, 120.
PAYTON III, D. N. and VISSCHER, W. M. (1966) *LA-3471-MS, UC-34, Physics, TID-4500*, Los Alamos Scientific Laboratory, University of California.
PHARIZEAU, P. and ZIMAN, J. M. (1963) *Phil. Mag.* **8**, 1487.
PIRENNE, J. (1958) *Physica* **24**, 73.
ROBERTS, A. P. and MAKINSON, R. E. B. (1962) *Proc. Phys. Soc.* **79**, 630.
ROBERTS, A. P. (1963) *Proc. Phys. Soc.* **81**, 416.
ROSENSTOCK, H. B. and MCGILL, R. E. (1962) *J. Math. Phys.* **3**, 200.
SAH, PRIYAMVADA and EISENSCHITZ, R. (1960) *Proc. Phys. Soc.* **75**, 700.
SAXON, D. S. and HUTNER, R. A. (1949) *Philips Res. Rep.* **4**, 81.
SCHLUP, W. A. (1963a) *Helvetica Phys. Acta* **36**, 41.
SCHLUP, W. A. (1963b) *Helvetica Phys. Acta* **36**, 570.

References

SCHLUP, W. A. (1965) *Phys. Kondens. Materie* **3**, 227.
SCHMIDT, H. (1957) *Phys. Rev.* **105**, 425.
TAKENO, S. (1960) *Prog. Theor. Phys.* **25**, 102.
TAKENO, S. (1962a) *Prog. Theor. Phys.* **28**, 33.
TAKENO, S. (1962b) *Prog. Theor. Phys.* **28**, 631.
TAKENO, S. (1962c) *Prog. Theor. Phys. Suppl.* **23**, 94.
TAKENO, S., KASHIWAMURA, S. and TERAMOTO, E. (1962) *Prog. Theor. Phys. Suppl.* **23**, 124.
TAKENO, S. (1963) *Prog. Theor. Phys.* **29**, 328.
TAYLOR, P. L. (1965) *Physics Letters* **18**, 13.
WADA, K. (1966) *Prog. Theor. Phys.* **36**, 726.
WALL, H. S. (1948) *Continued Fractions*, Van Nostrand Co., Princeton, 65.
YONEZAWA, F. (1964) *Prog. Theor. Phys.* **31**, 357.
YOUNG, C. and DWORIN, L. (1965) Unpublished.

Author Index

Author Index

Subject Index

Page references in italic figures denote items which have been emphasized in the text

Subject Index

Subject Index

Subject Index

MADE IN BRITAIN